D0856864

The Institute of Biology's
Studies in Biology No. 67

Animal Asymmetry

A. C. Neville
B.Sc., M.A., Ph.D.
Reader in Zoology, University of Bristol

Edward Arnold

First published 1976
by Edward Arnold (Publishers) Limited
25 Hill Street, London, W1X 8LL

Boards edition ISBN: 0 7131 2556 X
Paper edition ISBN: 0 7131 2557 8

Acknowledgements
   For bringing data to my attention I wish to sincerely thank the
following: Mr R. Abel, Dr W. B. Amos, Mr R. Campbell, Mr W. D. Clark,
Dr D. H. Godden, Dr R. J. Goldacre, Mr D. C. Gubb, Professor H. E.
Hinton, F.R.S., Mr J. Huxley, Dr P. J. Miller, Dr C. J. Pennycuick, Dr
T. E. Thompson, and Profesor Sir V. B. Wigglesworth, F.R.S. I also thank
Mr. D. F. Gilman who appears in Fig. 3–16, and Mr D. J. Roberts and
Mr. J. K. Wood for careful photographic assistance.

                                                                    A. C. N.

Printed in Great Britain by
The Camelot Press Ltd, Southampton

General Preface to the Series

It is no longer possible for one textbook to cover the whole field of Biology and to remain sufficiently up to date. At the same time teachers and students at school, college or university need to keep abreast of recent trends and know where the most significant developments are taking place.

To meet the need for this progressive approach the Institute of Biology has for some years sponsored this series of booklets dealing with subjects specially selected by a panel of editors. The enthusiastic acceptance of the series by teachers and students at school, college and university shows the usefulness of the books in providing a clear and up-to-date coverage of topics, particularly in areas of research and changing views.

Among features of the series are the attention given to methods, the inclusion of a selected list of books for further reading and, wherever possible, suggestions for practical work.

Readers' comments will be welcomed by the author or the Education Officer of the Institute.

1976   The Institute of Biology
                                              41 Queens Gate,
                                              London, SW7 5HU

Preface

This book concerns an initially simple concept. Certain animals show asymmetry; in some cases the asymmetry is apparent by comparison with symmetrical individuals of the same species, and in others when compared with symmetrical animals in related taxonomic groups. The asymmetry may be either structural (thereby posing interesting problems in developmental biology), or behavioural. The examples share a common theme: they cannot be mirror-imaged about the midline of an otherwise bilaterally symmetrical animal. Emphasis is given to the hierarchical nature of asymmetry: in one individual some levels of structure may be asymmetrical whereas others are not. The theme thus forms a thinking exercise in three dimensions. Wherever possible I have stressed the functional implications of different kinds of asymmetry. The examples chosen, many of which will be already familiar to sixth formers and undergraduates, are certainly far from exhaustive and readers will be able to think of many others for themselves.

I have thoroughly enjoyed getting together the material for this book. But its purpose is not entirely hedonic; I wish my readers pleasure in reading it.

Bristol,   A. C. N.
1976

Contents

1   Molecular Asymmetry   1
1.1 Isomer molecules in living systems   1.2 Helical macro-molecules   1.3 Helical fibrils   1.4 Conclusions

2   Types of Asymmetry in Organisms   6

3   Examples of Structural Asymmetry   8
3.1   Helical flagella in bacteria   3.2 An enantiomorphic protozoan   3.3 Asymmetry in coelenterates   3.4 Asymmetry in polychaete annelids   3.5 Heterochely in Crustacea   3.6 Hermit crabs   3.7 Asymmetry in parasitic arthropods   3.8 Wings overlap in insects   3.9 Miscellaneous asymmetry in arthropods   3.10 Asymmetry in molluscs   3.11 Echinoderm asymmetry   3.12 Asymmetry in fish   3.13 Asymmetry in birds 3.14 Asymmetry in humans   3.15 Muscular asymmetry 3.16 Pattern asymmetry

4   Asymmetry of Skeletal Fibres   26
4.1 Vertebrate eyes   4.2 Geodesic spirals and their functions 4.3 Helicoidal structures

5   Asymmetry in Development   36
5.1 Spiral cleavage in gastropods   5.2 Determination of asymmetry in snail eggs   5.3 Insect gynandromorphs   5.4 Labiopedia in insects   5.5 Asymmetry in bed bugs   5.6 Cell polarity gradients   5.7 Protochordate asymmetry   5.8 Gut asymmetry in Amphibia and fish monsters

6   Asymmetry in Behaviour   45
6.1 Asymmetry in arthropod behaviour   6.2 Asymmetry in mollusc behaviour   6.3 Ear asymmetry in owls   6.4 Asymmetry in marine mammals   6.5 Asymmetry in human brains 6.6 Spiral movements in man   6.7 Spiral movements in other animals   6.8 Functional uses of asymmetry

7   How is Asymmetry Achieved?   55
7.1 Bias in asymmetry   7.2 Enantiomorphic forms   7.3 Differentiation   7.4 Genetics and asymmetry   7.5 Cell gradients   7.6 Self-assembly   7.7 Conclusions

Appendix: Projects   59

References   60

1 Molecular Asymmetry

1.1 Isomer molecules in living systems

Many of the properties of living systems can be traced to the uniqueness of the element carbon. Since it has a valency of four, there exist two possible spatial configurations in compounds where the carbon is bonded to four different chemical groups. Such a carbon atom is known as *asymmetric*, and the two compounds are related in structure as is an object to its mirror image (Fig. 1–1). An asymmetric object has no plane of symmetry, i.e. it can never be made to coincide with its reflection in a mirror no matter how it is turned. A nice example to illustrate the

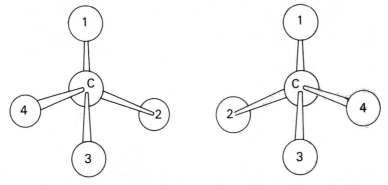

Fig. 1–1   Diagrams to show the two mirror-imaged ways in which four different groups may attach to an asymmetrical carbon atom.

difference between a symmetric and an asymmetric object is given by Martin Gardner, whose book *The Ambidextrous Universe* is highly recommended as introductory reading. Thus, the letter **B** appears upside down in a mirror, but appears correct when it is itself inverted. This is because it is a symmetric object with an axis of symmetry running horizontally across the letter. By contrast, the letter **F** is asymmetric and cannot be righted in a mirror: it has no axis of symmetry. An asymmetric object then has a reversed mirror image. The object and its image form an *enantiomorphic* pair. Examples are pairs of shoes and gloves, left and right-handed scissors, left and right individuals of the Protozoan *Zoothamnium* and of the fish *Anableps*, and of course the chemical compounds shown in Fig. 1–1. Such compounds are called *enantiomorphic isomers*. They possess *optical activity* in solution, being able to rotate plane polarized light in one case in a clockwise direction and in the other in an anticlockwise direction. The one isomer is *laevorotatory* (L) and the other *dextrorotatory* (D). They are defined by convention. (Plane polarized light is defined as light which vibrates in one plane only. It can be produced by passing ordinary light through a sheet of polarizing filter such as is used in polarizing sunglasses. The molecules in the filter are strongly oriented in one preferred direction so that light can only pass through if it vibrates in that direction. If such plane polarized light is viewed through a second polarizing filter it will black out when the preferred direction of the second filter is oriented at right angles to that of the first filter. Thus a horizontal source of plane polarized light would be cut out by a polarizing filter with its preferred direction placed vertically. If an optically active solution were placed between the polarizing filters, the observer would have to turn the filter nearest to him through an angle to re-establish black out, because the solution had rotated the plane of polarized light. By convention, if the filter has to be turned clockwise the substance is dextrorotatory; if anticlockwise, laevorotatory.)

Amino acids, which when polymerized in specific sequences constitute proteins, contain an asymmetric carbon atom (asterisked in Fig. 1–2). Here **R** is the chemical group which defines which amino acid it is. Sugars

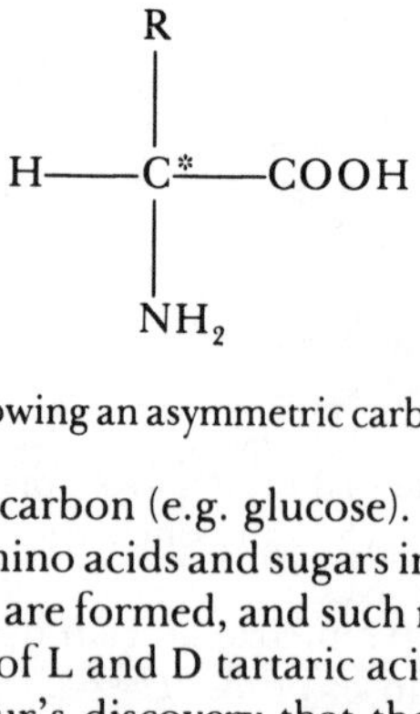

$$R$$
$$|$$
$$H\text{———}C^{*}\text{———}COOH$$
$$|$$
$$NH_2$$

Fig. 1–2 An amino acid showing an asymmetric carbon atom (asterisked).

also contain an asymmetric carbon (e.g. glucose). During total synthesis (i.e. not using enzymes) of amino acids and sugars in the laboratory, equal amounts of L and D isomers are formed, and such mixtures are known as *racemic* mixtures. A mixture of L and D tartaric acid used to be known as racemic acid prior to Pasteur's discovery that the two isomers formed differently shaped crystals which could be separated by hand and shown to form solutions of opposite optical activity. The formulae of three forms of tartaric acid are shown in Fig. 1–3. L–tartaric acid contains two left-handed asymmetrical carbon atoms whereas D-tartaric acid has two right-handed carbons. Mesotartaric acid has one of each so that they cancel to give no optical activity. L and D isomers occur naturally in 50/50 racemic mixtures unless produced by a living system. Laboratory prepared chemicals also occur as racemic mixtures unless asymmetry is applied during the synthesis, by using an existing asymmetric substance to inject left or right bias.

It is therefore highly interesting that living organisms contain only L-

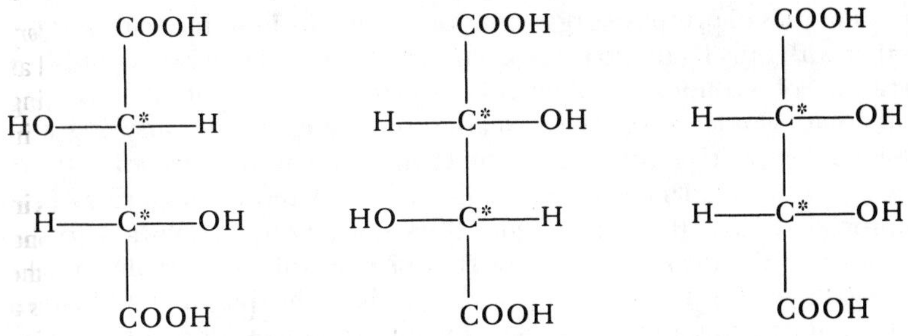

L-Tartaric acid   D-Tartaric acid   Mesotartaric acid

Fig. 1-3   Three forms of tartaric acid.

amino acids and D-sugars. Pasteur distinguished living from non-living systems on this basis. He showed that a fungus could select L-tartaric acid from a solution of L and D, so that the solution began to show optical activity. Therefore the fungus is asymmetric because it can separate enantiomorphs. All living systems are characteristically made of non-racemic mixtures of asymmetrical molecules. They are able to discriminate between L and D isomers taken in as a result of feeding on other living organisms which themselves contain asymmetrical molecules. On the human tongue, the taste receptor sites for sugars are asymmetric. If the left and right sides of the body were symmetrical the body would receive conflicting information, e.g. sucrose would taste sweet to one side of the tongue and not to the other side. Again, L-adrenalin has a twelve times greater effect on constricting blood vessels than does D-adrenalin, while the wrong isomer of vitamin C has no effect on the body at all.

How is such molecular asymmetry in living systems explained? There are two basic approaches to this fundamental question (which is of course of paramount importance in *theories of the origin of life*). Supporters of one camp propose that physical factors can adequately explain the problem. FRANK (1953) suggested that the type of isomer first formed could bias the whole of the rest of the synthetic process towards that isomer by competition. GARAY (1968) has shown that when racemic mixtures of the amino acids alanine, tryptophan and tyrosine in alkaline solution, are subjected to decomposition by radio-active β-particles of strontium[90] yttrium (in equilibrium), the D-isomers are destroyed more quickly than the L-isomers, leading to the selection of L-amino acids. (The β-particles from strontium[90], phosphorus[32] or potassium[40] decay are circularly polarized in one direction. See section 4.3 for an explanation of circular polarization.)

By contrast, another school of thought is that the choice of isomers occurred in biological systems themselves. AGENO (1972) imagines two populations of primitive organisms, one with only L-amino acids and the other with only D-amino acids, each having selected one isomer only for reasons of economy of metabolic reactions. Such populations with different symmetry would be unable to produce fertile offspring and eventually one type survived by the chances of natural selection. Of the two types of explanation that of Garay seems more convincing. It is unfortunate that thermodynamics does not provide a solution to the problem, since the kinetics of L or D or racemic solutions are the same.

Again one can ask why living organisms have the specific combination of L-amino acids and D-sugars. Why not L-amino acids and L-sugars? Here, molecular wire models help to provide a solution. If you build an L-amino acid you can convert it to a D-sugar with far less rearrangement than if you try to convert it into an L-sugar. (It is worth obtaining a molecular model kit and making model isomers.)

A meteorite which fell recently near Murchison in Australia was found to contain racemic mixtures of L and D-amino acids. It is clear that the basic chemical requirements for life do not exist on our planet alone.

1.2 Helical macromolecules

When chains of amino acids form an α-helix, the most stable sense of helix which can be formed by L-amino acids is a right-handed one (i.e. with the same twist as a right-handed corkscrew: left-handed corkscrews are specially made for left-handed people). The right-handed twist is favoured because the hydrogen bonds formed between different turns of the helix are then stronger. You can verify this with molecular model kits. If your build an L-α-helix with L-amino acids, building in hydrogen bonds of optimal length with pieces of wire, you will find that they are distorted. In life, all α-helices are right-handed. The sense of twist is known as the *chirality* of the helix. Many biological macromolecules are helical, at least in part, and because of similar stability arguments they exist with only one sense of chirality, e.g. nucleic acids, keratin, actin, myosin, starch, collagen. A helix has the same sense of chirality no matter from which end you look at it. (Try this with a wire helix.) A mirror reverses the sense of a helix. Since biological helices have the same sense on both sides of the body, they too represent a type of structural asymmetry.

1.3 Helical fibrils

Molecular asymmetry gives rise to crystalline aggregates of much larger size which also show asymmetry. A nice illustration of this is given by Tachibana and Kambara (1967) for the polymer polybenzyl glutamate. This may exist in a D-form (PBDG) or an L-form (PBLG). The two isomers form D and L-helically twisted fibrils respectively (Fig. 1–4) and these authors give electron micrographs illustrating this difference. The fibrils are from 0.1 to 1.0 μm wide. Interestingly, racemic mixtures of PBDG and PBLG gave non-helical fibrils. There is a change in sense at each level of structural unit. Thus the PBLG forms D-α-helices which form superhelical fibrils with an L-handed twist. As in rope, the sense of a

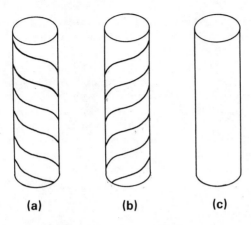

(a)   **(b)**   **(c)**

Fig. 1–4 Fibrils of polybenzylglutamate. (a) PBLG forms a left-handed helical fibril. (b) PBDG forms a right-handed one. (c) A racemic mixture of PBLG and PBDG forms non-helical fibrils.

superhelix is opposite to that of the component helices, as suggested by Crick (1953). Convince yourself that different hierarchical levels of helical component pack together better when the chirality alternates, by stripping down a piece of rope.

1.4   Conclusions

It follows that if all living organisms contain only L-amino acids, then at this level of analysis there can be no such thing as a bilaterally symmetrical animal. Such animals are really *pseudosymmetrical*. Again, we have seen that biological helices have one sense of twist which is the same on both sides of the body. The problem then is how to build an apparent bilateral symmetry with asymmetrical components. Furthermore, how is asymmetry achieved at higher levels of structure? This leads us to a consideration of the hierarchical nature of asymmetry which is the subject of the next chapter.

2 Types of Asymmetry in Organisms

Bilaterally symmetrical animals can be divided by a sagittal cut into two mirror-imaged halves. Asymmetry may occur in such an animal in several ways. (a) As we have seen in the previous chapter, all organisms are asymmetrical at the molecular and macromolecular levels. (b) Its external structure as viewed from the outside, may become asymmetrical (Fig. 2–1a, b). Typical examples are lobsters with the crushing claw much larger on one side; flatfish lying always on one side; colour pattern in the common viper. (c) Internal structure may be asymmetrical even though bilateral symmetry is retained at the whole animal level. Examples are the heart and arterial arches in man (Fig. 2–1c) and the fibre orientations in

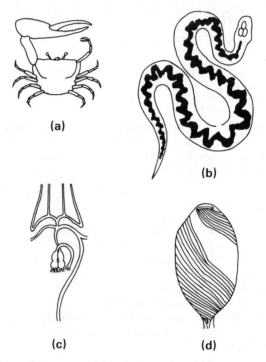

(a)

(b)

(c)

(d)

Fig. 2–1 Types of Asymmetry. (a) At the appendage level, in a male fiddler crab. (b) At the surface pattern level in a viper. (c) At the internal organ level, in the aortic arches of a mammal. (d) At the fibre level; some of the fibres in a layer of basement membrane from a tadpole, *Bombinator*.

the basement membrane of tadpoles (Fig. 2–1d). The origin of such asymmetry can be traced during development (Chapter 5). Symmetry of external form is often retained as an adaptation to rapid movement by streamlining, whilst internal organs can more readily become asymmetrical. (d) Behaviour may be asymmetrical (Chapter 6). What we are *not* concerned with in this book are the more trivial examples of animals which are shapeless (e.g. *Amoeba proteus*) and hence always asymmetrical.

Biological structure is hierarchical, i.e. there are many levels of structure (molecules, macromolecules, microfibrils, fibrils, fibres, cells, organs, whole organism). It follows that whereas some levels of structure may be bilaterally symmetrical, others may be asymmetrical. For example, as we have seen in Chapter 1, all organisms are asymmetrical at the molecular level since they contain only L-amino acids and D-sugars. Also, at the macromolecular level, any protein which contains α-helical regions will have them arranged as right-handed helices on both sides of the body. Yet bilateral symmetry may prevail at other levels, for example general body shape. In adult beetles, the orientation of the first layer of cuticle laid down after hatching follows the bilateral symmetry of the body. But subsequent layers are laid down with asymmetrical orientation (see Chapter 4). It is clear that there must exist separate control mechanisms for the various types of asymmetry. An attempt to approach this problem is made in Chapter 7. But before discussing mechanisms we must present a range of telling examples of asymmetry and this is the purpose of the next few chapters.

3 Examples of Structural Asymmetry

3.1 Helical flagella in bacteria

Bundles of bacterial flagella containing many turns of left-handed spirals will not entangle provided that they rotate clockwise when viewed from the body. In this case points of entanglement travel outwards towards the free ends of the flagella and the entangled flagella then, become free. If rotation was anticlockwise such travelling waves of entanglement would jam when they reached the body surface. Hence rotation occurs in one sense only so that both structure and motion of the flagella are asymmetrical with respect to the whole bacterium (Fig. 3–1).

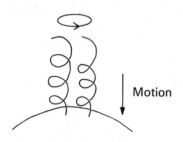

Motion

Fig. 3–1 The relation between sense of helix and sense of rotation in bacterial flagella. (After LOWY and SPENCER (1968), *Symp. Soc. exp. Biol.*, **22**, 215–37.)

3.2 An enantiomorphic protozoan

Zoothamnium arbuscula is a remarkable colonial Protozoan with a characteristic branching habit and a contractile stalk. The species exists in two enantiomorphic forms with dextral and laeval branching patterns which are related as mirror images (Fig. 3–2). Out of 50 colonies examined, 27 were dextral and 23 laeval.

3.3 Asymmetry in coelenterates

Some colonial Coelenterates float on the surface of the sea, being blown along by the action of the wind on a special structure which acts as a sail. In two species which are sometimes washed ashore in the British Isles, the Portuguese Man-o'-war (*Physalia physalis*) and the Jack Sail-by-the-wind (*Velella velella*), the sail is set at an angle to the body axis. The progeny of one parent are polymorphic, some individuals having the sail from right to left, others from left to right. They are thus blown in opposite directions, reducing the risk of all the offspring being washed ashore.

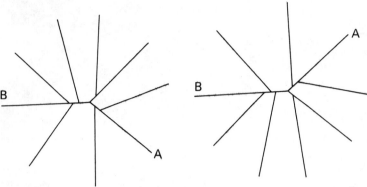

Fig. 3–2   Left and right branching *Zoothamnium arbuscula*, in ventral view. The two colonies are mirror-images of each other. (After FURSSENKO (1929), Archiv. *Protistenk*, **67**, 376–500.)

3.4   Asymmetry in polychaete annelids

Associated with a free-swimming life, there is almost no asymmetry in the external form of errant polychaete worms. However, the pharyngeal apparatus is often asymmetrical, since the maxillary plate on one side enters a space between two plates on the other side (e.g. Eunicidae). By contrast, asymmetry is found in sessile forms, e.g. the spiral tentacular apparatus in *Spirographis*. Also there is an asymmetrical operculum in Serpulidae. In *Hydroides* the paired opercula consist of one which is large and functional and one which is rudimentary. Amputation of the large operculum causes the rudimentary one to grow and become functional. The amputated one regenerates but is now of rudimentary type. There are parallels in the regeneration of heterochelous claws in Crustacea (see section 3.5).

3.5   Heterochely in Crustacea

Heterochely is the unequal development of the chelae (claws) in certain (Decapod) Crustacea. For example, a lobster has a heavy claw on one side of the body which is adapted for crushing. Its homologue on the other side of the body is adapted as a light nipping claw for picking up food. Heterochely reaches a climax in the male fiddler crab, *Uca pugnax*, in which the large claw can be up to twenty times heavier than its homologue, representing up to 70% of total body weight (Fig. 3.3). This degree of heterochely is not found in swimming Decapod species which would be overbalanced. In some crabs which have secondarily forsaken bottom-dwelling for a free-swimming existence, heterochely may persist in *size* but not in *weight*.

The fossil record shows that heterochely is an advanced character,

Fig. 3–3 Female and male fiddler crabs (*Uca pugnax*). The male has a much enlarged claw.

usually restricted to males but found in both sexes in several species. Some females of *Xantho* have normal female sex organs combined with a male type crushing claw.

Crustacean heterochely is determined in each individual by the chance loss of an appendage (they are adapted to autotomize in escape situations) followed by regeneration through several moulting cycles. In *Uca* only males develop the large chela, which is used in territorial display and fighting. *Uca musica* stridulates by rubbing the enlarged chela against the first walking leg on the same side. The necessary ridges are present only on one side of the animal. The stridulation gives rise to the name fiddler crab. Young males have two equal chelae, both of male type. Subsequent growth depends upon their experiences. If one chela is lost, it regenerates as a small female type chela. Thus heterochely is normally 50% left or right handed since it is determined by chance injury in young crabs. If both chelae are lost then two small female type ones are regenerated. If, however, neither chela is lost, the adult ends up with two symmetrical male type chelae. In *Uca*, once regeneration has occurred, the decision is final: if subsequently lost again, a regenerated chela will only be replaced by one of its own type. (Interestingly, these discoveries were made by Morgan, the man responsible for mapping the position of genes on chromosomes in *Drosophila*.)

The determination of chela type during regeneration is more complex in some other Crustacea. Thus Przibram (1930) found that if a small female type chela is amputated in the snapping shrimp *Alphaeus*, the same type is regenerated. If, however, a male chela was removed, a reversal occurred, with the surviving chela developing into a male type. This also happens in the lobster *Homarus*, but only if the experiment is performed early enough in the lifetime. By contrast, in *Alphaeus* reversal can occur throughout life. If both chelae are autotomized, the one on the original nipper side will become a crusher if it is given about forty hours start over the other claw. Normal heterochely (i.e. without reversal) occurs, however, if both claws are removed simultaneously. There is a competition between claws.

It was heterochely in male fiddler crabs which first inspired Sir Julian Huxley to work on problems of *allometric growth*. Allometry occurs when a part of the body grows at a different rate from that of the body as a whole. It is immortalized in Huxley's formula,

$$y = bx^a$$

where $y =$ size of an organ; $b =$ original size of the organ; $x =$ size of the rest of the body without the organ; $a =$ percentage rate of growth of the organ. When $a = 1$ then the relative size of the organ with respect to the body remains constant (e.g. the small female type claw of a male fiddler crab). However, when a is greater than unity, the organ becomes relatively larger (e.g. the huge chela of the male fiddler crab, which increases from 2% to 70% of the weight of the rest of the body).

3.6 Hermit crabs

These show a nice series of degrees of asymmetry, related to habitat (Fig. 3–4). They arose as primarily symmetrical forms (e.g. Pylochelidae,

Fig. 3–4 Land hermit crab (*Coenobita*).

Pomatochelidae) living in the empty symmetrical shell of *Dentalium* (Mollusca: Scaphopoda) or in sunken pieces of bamboo. The family Paguridae have developed striking asymmetry in relation to living in empty shells of dextrally coiled Gastropod Mollusca. They grip the columella of the host shell with the enlarged left uropod, and have specially soft abdominal cuticle for gripping the shell by hydrostatic pressure from within. The abdomen is asymmetrically curved, with asymmetrical tergites, and in females the second to fifth pleopods are absent on the right-hand side. The right chela is much the larger and is used to guard the entrance to the host shell.

The robber crab, *Birgus* (Coenobitidae) leaves the sea with a symmetrical abdomen at the completion of metamorphosis from the larval stages. It then spends a few months inhabiting a dextral gastropod shell out on the land, during which time the abdomen becomes asymmetrical. It then outgrows its shell, develops lungs and becomes free living on land, and becomes symmetrical once more. The Lithodidae become secondarily free-living, in this case retaining the asymmetrical abdominal tergites.

Brightwell (1952) found that if a hermit crab was presented with a glass model of a sinistral mollusc shell, it rejected it immediately in favour of a dextral one. One French scientist once took the long spires of dextral *Turritella* shells and cut off the apex, believing that he had created a sinistral opening. When hermit crabs were shown to accept such openings he thought that he had proved that hermit crabs could accept sinistral homes. But of course he was wrong (we have seen in Chapter 1 that a right-handed corkscrew twists the same way no matter from which end you look at it). Why do you think there are no sinistral hermit crabs?

In *Eupagurus prideauxi* asymmetry appears during larval development and seems to be specifically dextral by the onset of post-larval life. This is presumably in adaptation to the dominance of dextral forms in the mollusc shells which they later inhabit. The greatest asymmetry is between the right and left chelae of the males, the right being dominant. This right-handed dominance decreases in appendages lying both anterior and posterior to the chelae, with a reversal of dominance occurring in the fifth pereiopod and uropods (in both of which the left is larger). There is less right-handed dominance in females, with the reversal point lying further forwards. A similar unilateral gradation of dominance exists in the male fiddler crab, *Uca*, though here it is set up by chance loss of a chela in early life.

3.7 Asymmetry in parasitic arthropods

The Isopoda contains several examples of parasites which have become asymmetrical in adaptation to their habitat. In *Bopyrus*, which is a parasite of prawn and shrimp gill chambers, mature females show asymmetry corresponding to which side of the host they inhabit. This is said to result from the unequal pressure of the walls of the gill chamber. Similar asymmetry is seen in *Phryxus* which lives in the gill chambers of Decapoda. A nice sequence of asymmetrical animals is shown by female *Atheges paguri* which are parasitic upon the asymmetrical abdomens of hermit crabs, which in turn live in asymmetrical mollusc shells.

In the copepod *Botryllophilus*, which is a parasite of the gill cavity of Ascidians, the thoracic limbs are asymmetrical. One side is adapted for swimming, the other side being hook shaped in adaptation to life in the confined space of the host's gill cavity.

Some feather mites (Acariformes: Analgesoidea) show asymmetry of the front legs, setae, rear of body, and body form as an adaptation to living in the asymmetrical spaces between the barbs and barbules of bird feathers.

3.8   Wing overlap in insects

Those insects which settle with their wings overlapping one over the other (e.g. many moths, cockroaches, grasshoppers, bugs, crickets, mantids) often do so in a specific sequence. For example, in Tettigonioidea the front wings usually overlap left over right (L/R), whereas in Gryllodea it is usually R/L. In Tettigoniids the bias in wing overlap is accompanied by asymmetry in the stridulatory apparatus, as is shown in the photographs of the wing bases of the British great green grasshopper, *Tettigonia viridissima* (Fig. 3–5). One forewing specializes as a scraper or plectrum, combined with a resonating area, the mirror. When the scraper is rubbed along a toothed file on the other wing it sets the mirror vibrating and a song is produced. If the wings fold L/R, the two wing surfaces which will tend to specialize in stridulation are the ventral side of the left forewing and the dorsal side of the right forewing.

The wings in the cricket *Gryllus bimaculatus* normally rest R/L. If they are experimentally folded the wrong way round after cuticular hardening has taken place, they usually return to R/L within a few minutes. If, however,

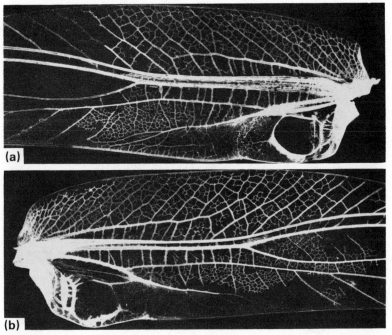

Fig. 3–5 (a) Left and (b) right forewing bases of great green grasshopper (*Tettigonia viridissima*). The wings were used as photographic negatives. (Photographs by Mr J. K. Wood.)

this is performed soon after adult emergence (before hardening has proceeded very far) they are unable to correct this, and later produce a weak courtship song. Also, they are unable to produce the fighting song at all. Perhaps this explains the low percentage of natural errors in wing folding (Table 4).

Acheta veletis normally stridulates R/L with the left scraper acting on the right file. There is only a slight difference in tooth number between left and right files. When the left scraper was removed, the crickets could not initially stridulate. However, 80% subsequently overcame their disability by reversing their wing folding, and using the right scraper instead.

Scheie and Smyth (1972) have recently found a remarkable active membrane response in the integument of a cockroach, *Supella supellectilium*. 100 millivolts electrical depolarization led to an active further depolarization of 50 millivolts. The response is probably located in the dermal glands. In the forewings these are only numerous on those parts which are not covered by the normal overlap of the forewings (L/R). Their distribution is therefore asymmetrical and the active electrical response shows similar asymmetry.

3.9   Miscellaneous asymmetry in arthropods

Free-swimming copepods swim with their antennae. In males the last pair of legs are used in copulation and often show asymmetry, as do the antennae which are modified for clasping the female. Such asymmetries are sometimes compensated by asymmetrical outgrowths on the sides of the last thoracic segment and the first abdominal segment, or by asymmetry of the caudal fork. Balance is thus restored so that they can maintain a straight course in swimming.

Ostracods often show asymmetrical shell valves, and among insects, mandibles often cross asymmetrically though I have not seen any figures of possible bias to one side occupying the top position. In the beetle *Xylotrupes gideon* the mandibles are asymmetrically shaped (Fig. 3–6): this feature can also be seen in silk moth caterpillars and adult locusts. A characteristic of the thrips (Thysanoptera) is asymmetry of the head capsule (Fig. 3–7).

Fig. 3–6   Asymmetrical mandibles of a beetle, *Xylotrupes gideon*.

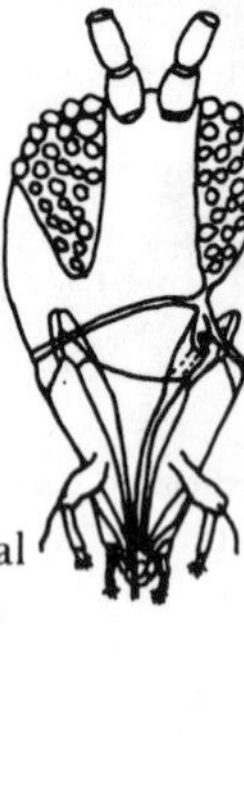

Fig. 3–7   A characteristically asymmetrical head in a thrip, *Aelothrips fasciata*.

3.10 Asymmetry in molluscs

The development of asymmetry is the major evolutionary trend of gastropod molluscs. The asymmetry is obvious even by casual inspection of a shell. The majority of gastropods have dextrally coiled shells, that is they twist clockwise as seen looking down at the shell apex. There are some species, however, which twist the other way (sinistral). A number of these are British freshwater snails (e.g. *Ancylus rivularis*, *Physa heterostropha*, *Physa fontinalis*, *Physa hypnorum*, *Planorbis marginatus*, *Planorbis carinatus*, *Planorbis trivolvis*, *Aplexa* spp.). Others are found in tree crevices (e.g. *Balea perversa*, *Marpessa laminata*, *Clausilia bidentata*). A quick way to tell if a species is dextral or sinistral is from the way in which the shell is carried (Fig. 3–8). Those who live near the outcrops of the Red Crag (Pleistocene) on the east coast of England, will be able to collect the large sinistral fossil species, *Neptunea contraria*.

The obvious distinction between gastropod and bivalve molluscs is that the former have one shell valve whereas the latter have two. There is, however, a remarkable gastropod which has secondarily evolved a second valve: this is *Tamanovalva limax* from the Inland Sea of Japan. The valve

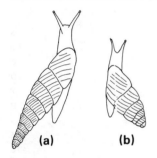

Fig. 3–8 Diagrams to show a quick way to distinguish sinistral from dextral snails by the manner of carrying the shell. (a) *Marpessa laminata* (sinistral). (b) *Ena obscura* (dextral).

which is the usual gastropod shell has a small sinistral spiral at the apex, even though the rest of the animal's viscera are arranged dextrally.

Occasionally a sinistral variety occurs in a species which is normally dextral. This is illustrated for *Cepaea nemoralis* (Fig. 3–9). Such examples are richly prized by collectors and are sold for high prices. Their rarity is indicated by the data in Table 4. However, in the Cuban tree snail, *Liguus poeyanus*, dextral and sinistral individuals occur in roughly equal numbers. Examples of *Limnaea peregra* have been bred in genetical experiments in which the left and right twisting forces appear to be balanced. The result is a shell which is coiled in one plane (BOYCOTT *et al.*, 1931).

Asymmetry also occurs in gastropod viscera. There is a reduction in dextral species of the right lobe of the liver and the right ctenidium. The right osphradium is lost and there is an evolutionary trend towards loss of the right auricle of the heart.

Fig. 3–9 Sinistral (left) and dextral (right) examples of a snail (*Cepaea nemoralis*).

Although on casual inspection lamellibranchs (bivalves) appear bilaterally symmetrical, the shell valves are in fact asymmetrical at the hinge since the teeth interdigitate. Species which live attached to the substratum develop asymmetrical shell valves; you can see this the next time you eat an oyster. In *Anomia*, the attached valve has a hole through which passes the byssus fastening the animal to the substratum. Inequal valves *par excellence* occur in the Mesozoic fossil group Rudistes. For example, *Hippurites goganiensis* had a tall conical right valve attached to the substratum, and a flat left valve arranged as a lid, so that it superficially resembled a solitary coral (Fig. 3–10).

Fig. 3–10 *Hippurites goganiensis*, a highly asymmetrical fossil lamellibranch. The small left valve acts as a lid for the large right valve.

Fig. 3–11 *Nipponites*, an asymmetrical ammonite fossil.

Among the cephalopods, the squids are mostly adapted to a fast swimming life and are bilaterally symmetrical for streamlining purposes. However, *Calliteuthis reversa* is a small deep water squid with asymmetrical eyes. The right eye is smaller and surrounded by a ring of photophores which emit light; the left eye lacks the ring of light organs. Some fossil ammonites had an asymmetrical shell (Fig. 3–11).

3.11 Echinoderm asymmetry

This group, famous for its basic five-fold symmetry, probably had bilateral symmetry early in its evolution. The larvae have the gut thrown into a spiral loop and there is reduction of the right coelom. The single madreporite and stone-canal are also asymmetrically positioned. Ohshima (1921) described mirror-image pluteus larvae of *Echinus miliaris* in which coelom, stone-canal, axial sinus and madreporic vesicle all showed *situs inversus*.

3.12   Asymmetry in fish

Most species of flatfish (Pleuronectidae) undergo a striking metamorphosis some time after hatching from the egg, becoming so asymmetrical that they spend the rest of their lifespan lying on one side only. In the plaice (*Pleuronectes platessa*), the young fish is symmetrical for up to 30 days. It then sinks to the bottom and lies on its left side, twisting the head through as much as 70 degrees so as to be able to see above itself with the lower (left) eye as well as the right one. This twisting occurs whilst the skull is still cartilaginous and flexible. Then the left eye begins to migrate upwards and forwards by growth. At 40 days old the left eye appears on the upper margin of the head anterior to the right eye, and by 45 days is in position above and in front of the right eye. In general, in species which come to lie on their left side the eyes come to lie on the right side, and vice versa. Metamorphosis involves several other changes. There is a change in behaviour from pelagic to bottom-living. Sometimes the jaws become asymmetrical. The pigmentation disappears from the blind underside, but becomes cryptically coloured for concealment on the upper side. When disturbed, the fish now swims with the coloured side uppermost.

The eyed side in flatfish has a better developed lateral line sense organ, and its paired fins are usually larger and with more fin rays than the blind side. In addition the eyed side often has ctenoid scales whereas the blind side has cycloid scales. The flatfishes are the most asymmetrical of all vertebrates. A most useful review of fish asymmetry is given by HUBBS and HUBBS (1945).

British flatfish fall into four groups: (1) Plaice, lemon sole, witch, dab and flounder form a group (Pleuronectidae) with the eyes on the right and teeth better developed on the blind side. (2) Halibut and long rough dab (Fig. 3–12) (Pleuronectidae) also have the eyes on the right but have symmetrical jaws. (3) Sole, sand sole, thick back and solenette (Soleidae) also have the eyes on the right but with teeth only on the left-hand side. (4) Turbot, megrim (Fig. 3–13), scaldfish, common topknot, Bloch's topknot and Norwegian topknot (Bothidae) have the eyes on the left-hand side, with symmetrical teeth and jaws.

Odd exceptions occur with the underside wholly or partly coloured. Individuals with the eyes on the wrong side are very exceptional, being less than one in 100 000 for sole and plaice. In flounder reversed specimens are more common. Primitive flatfish, e.g. the Indian genus *Psettodes* has about half the individuals left handed and half right handed. There seems to be an evolutionary trend towards greater accuracy of bias.

18

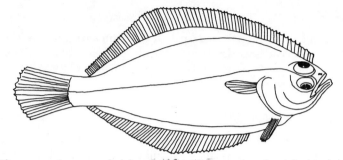

Fig. 3–12   Long rough dab, a flatfish with the eyes on the right-hand side.

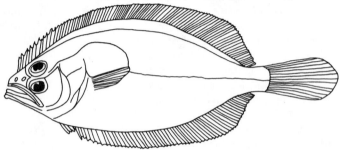

Fig. 3–13   Megrim, a flatfish with the eyes on the left-hand side.

Hippoglossus pinguis remains symmetrical throughout life. Notice that the bony flatfishes mentioned above have adopted asymmetry in adaptation to bottom-dwelling. By contrast, the cartilaginous bottom-living flatfish (skates and rays) have retained bilateral symmetry.

A remarkable case of asymmetry is found in some of the viviparous tooth-carps of South America. In the four-eyed fish *Anableps anableps* (Anablepidae) and the one-sided live-bearer *Jenynsia lineata* (Jenynsiidae), males have the anal fin modified for use as an intromittent organ (gonopodium). This acts as a tubular extension of the sperm channel and can only be moved, by the aid of special muscles, to one side (Fig. 3–14a). Moreover, females have the sex opening on one side also, due to partial occlusion by a large scale (Fig. 3–14b). Hence a male with a right-moving gonopodium can only mate with a female whose opening is on the left.

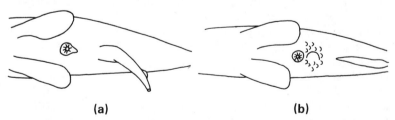

(a)   **(b)**

Fig. 3–14   Ventral views of the four-eyed fish, *Anableps anableps*. (a) Dextral male. (b) Sinistral female.

This is thought to have evolved from the habit of these fish to swim side by side in pairs. Garman (1895) who discovered this asymmetry, found that in *Anableps* one part of the breeding population comprises 60% dextral males and 60% sinistral females, the other part comprising 40% sinistral males and 40% dextral females. In *Jenynsia lineata* the wild population consists mostly of dextral males and sinistral females.

Males of Phallostethidae are also structurally left or right individuals and have an asymmetrical mating behaviour. In *Horaichthys setnai* the female lacks the right pelvic fin, providing a space for the attachment of the spiny spermatophore from the male.

Other examples of fish asymmetry include hagfish gill slits, sawfish rostral teeth and Cyprinid pharyngeal teeth.

3.13   Asymmetry in birds

The only bird with a beak which bends to one side is the wry-billed plover (*Anarhyncus frontalis*), found on the shingle-river beds of Canterbury, New Zealand. The beak is one and a quarter inches long, and is always bent to the right-hand side in the middle so that the end is 12 degrees offset from the base (Fig. 3–15). It is used for turning over stones when searching for food, and is thus an advantage in searching on the right-hand side but a hindrance on the left-hand side. This influences their feeding behaviour.

The crossed beak of the crossbill (*Loxia curvirostra*) is specially adapted for picking seeds out of fir cones. The beak tips may cross in either

Fig. 3–15   The New Zealand wry-billed plover, *Anarhyncus frontalis*, showing the asymmetrical beak.

direction. A recent study in Norway showed that the lower mandible crossed to the left in 58% and to the right in 42% of examples, independent of sex. Young crossbills are born with straight bills which lengthen around day 14 but are still uncrossed when they fledge at 20 days. Crossing starts at around day 30 but they are unable to feed themselves until about day 45.

Domestic fowls are sometimes born with an extra hind digit on one side (heterodactyl). When four-toed breeds were crossed with five-toed breeds, 38 out of 402 offspring were heterodactylous, with four toes on one foot and five on the other. Nearly 90% had the extra digit on the left-hand side. However, in the first generation resulting from crossing such heterodactylous fowls, the extra digit was always on the left. Only in the second generation did a few right-hand sided examples appear. In further experiments, heterodactylous fowls were bred with six toes on one foot and five on the other.

When fish tailed and fan tailed pigeons are crossed (they have different numbers of tail feathers) the resulting hybrids have a slight asymmetry in number of tail feathers, with a tendency to slight bias to one side.

3.14   Asymmetry in humans

Venus de Milo has an asymmetrical head. This reflects the accuracy of the sculptor, since cranial asymmetry is a human characteristic. The only other Primates with asymmetrical skulls are gorillas. There is some discrepancy in the literature as to which side of the human skull is larger. Measurements on ancient Egyptian skulls from the 26th to the 30th dynasties showed the right side of the cranium to be larger. However, Vig and Hewitt (1974), investigating a sample of skulls of 60 normal children by radiography, found that the majority had the left side larger. The maxillary region was, however, larger on the right-hand side and they suggest that this compensates for the upper cranial asymmetry to give a correct chewing action. Mulick (1965), however, found that the face was larger on the left-hand side in six same-sex triplets. Also, the ancient Egyptian sample had a larger zygomatic arch and maxilla on the left-hand side.

Right-handed people develop larger right hands (glove manufacturers allow for this since about 95% of people are right-handed). Also the right arm tends to be longer.

The greater length of the left side of the skull between temporal fossa and gnathion in *Gorilla* has been suggested to be a result of asymmetrical mastication.

That the human face is asymmetrical can be shown by photographic reconstruction of a symmetrical face from either two left or two right halves (Fig. 3–16). It is difficult to identify an individual thus reconstructed. This is presumably a fault common to many police 'Identikit' pictures.

In Siamese twins there is always a reversal of one set of organs in one twin (*situs inversus*), e.g. the heart is changed to the right and the liver to the left-hand side. We thus reach the surprising conclusion that Siamese twins are *not* identical! They are like book-ends (enantiomorphs). This sometimes occurs in ordinary twins. Tenniel's illustrations to *Through the Looking-glass* by Lewis Carroll, show the twins Tweedledum and

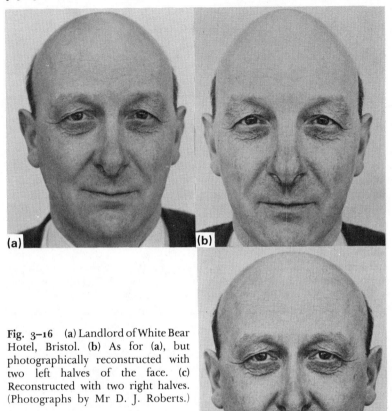

Fig. 3–16 (a) Landlord of White Bear Hotel, Bristol. (b) As for (a), but photographically reconstructed with two left halves of the face. (c) Reconstructed with two right halves. (Photographs by Mr D. J. Roberts.)

Tweedledee as a mirror-imaged pair, showing his accuracy of observation.

The human umbilical cord is usually a left-handed triple helix of two veins and one artery.

3.15   Muscular asymmetry

A number of asymmetrical structures owe their origin to the contraction of asymmetrically balanced sets of muscles. For instance, in the hermit crab, *Pagurus granosimanus*, the central abdominal flexor muscles of the right-hand side are three times the diameter of those on the left. Thus their contraction causes the abdomen to curl so as to match the twist of the dextral gastropod shell in which the crab lives.

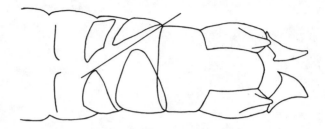

Fig. 3–17 Ventral view of abdomen of a male *Clunio marinus* after torsion. (After DORDEL (1973), *Z. Morph. Tiere*, **75**, 165–221.)

In many flies, copulation can only occur after the body of the male has undergone a muscular torsion. In the blowfly *Calliphora* the torsion occurs during the early pupal stage, whereas in *Clunio* (a small fly which is common on washed-up seaweed) it occurs during the first two hours of adult life. This torsion is caused by the extensive contraction (up to 50% of their length) of special muscles which run helically around the posterior part of the abdomen. Each of the three abdominal segments 6, 7 and 8, turns through 60 degrees with respect to its neighbour, to give a combined torsion of 180 degrees (Figs. 3–17 and 3–18). In *Clunio*, as in the mosquitoes *Culex* and *Aedes* this torsion is permanent. In *Tipula paludosa* it is temporary. Even within a single genus this characteristic may vary, e.g. in *Chironomus plumosus* it is permanent whereas in *C. nuditarsis* it is temporary. Besides making mating possible in *Clunio*, muscular torsion also shortens the body, thus increasing its skeletal tubular stability and also moving the centre of gravity forwards. The male is then better able to

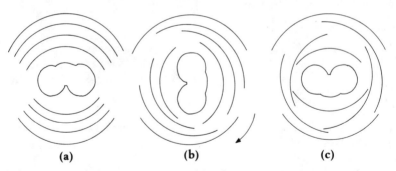

(a)	(b)	(c)

Fig. 3–18 Diagrammatic representation of torsion in the abdomen of a male *Clunio marinus* to show the relative changes in position of the sclerites. (a) Before torsion. (b) Torsion half-completed. (c) Torsion completed. (After DORDEL (1973), *Z. Morph. Tiere*, **75**, 165–221.)

carry the wingless female during the mating flight. Richards (1927) records that the torsion is from left to right in *Bombylius discolor* and the opposite way in *Volucella pellucens* and *Calliphora*. However, it may well be that a large sample would show a mixture of both, with the bias to favour one sense of twist changing with geographical habitat as Dordel (1973) has found for *Clunio* (Table 5). The degree of asymmetry which is found in some flies is well illustrated by the genitalia of a male hover fly, *Chrysotoxum cautum* (Fig. 3–19). The genitalia of many insects are asymmetrical and this helps in the reproductive isolation of each species (by analogy with a lock and key) as well as giving rise to special mating postures (e.g. in the hawkmoth *Hemaris* pairing takes place side by side). Burns (1970) notes that whereas the male genitalia of the moth *Erynnis* are usually highly asymmetrical, occasional individuals occur with secondary mirror-imaged symmetry.

Fig. 3–19 Asymmetrical genitalia in a male hover fly (*Chrysotoxum cautum*). (Photograph by Mr J. K. Wood.)

Muscular torsion is one of the characteristic features of gastropod molluscs. There is a change at metamorphosis from ciliary locomotion in the larva to muscular locomotion in the adult. The cilia occur on a velum which has a muscle for its retraction. As a result of spiral cleavage (see Chapter 5) a band of mesoderm develops from a large cell, 4d (Fig. 5–4). This cell occupies a mirror-imaged position according to whether the snail is dextral or sinistral. In a dextral snail it gives rise to the retractor muscle of the right side. There is no corresponding left retractor. Contraction of the muscle puts an asymmetrical strain on the developing tissues, causing a rotation of 180 degrees of the visceral hump and shell with respect to the head and foot. In a dextral snail the rotation is anti-clockwise as viewed from the dorsal aspect, i.e. in the same sense as the coiling of the shell. Torsion is, however, a separate process from the spiral growth of the shell: often torsion occurs before any shell spirals are visible. Torsion in *Fissurella* can occur as quickly as two to three minutes in *Acmaea*. However, in *Haliotis* there is an initial muscular phase of torsion through about 90 degrees which takes 3 hours, followed by a slower growth phase lasting 200 hours. In *Pomatias elegans*, which is a land snail, there is no larva. The velar retractor muscles have been lost and torsion occurs entirely by asymmetrical growth over a 10 day period (Creek, 1951).

3.16   Pattern asymmetry

Chromatocyte cells containing pigment are visible through the transparent larval cuticle of a fly, *Thaumalea verralli*. The colour pattern which they form is bilaterally asymmetrical. This helps to camouflage the larva by breaking up its outline better than would a symmetrical pattern. Furthermore, the pattern differs from segment to segment making it less conspicuously repetitive. It is surprising that insects have not made wider use of asymmetrical patterns for improved concealment.

In the butterfly *Urania ripheus* from Madagascar, beautiful interference colours are produced by thin layers of cuticle in the scales. Casual inspection gives the impression that the pattern on the wings is bilaterally symmetrical, but more detailed examination (see Fig. 3–20) reveals considerable pattern asymmetry. The specimen shown is from my own collection (and I am indebted to Professor H. E. Hinton, F.R.S. for first

Fig. 3–20 *Urania ripheus*. Spot the differences between left and right halves. (Photograph by Mr J. K. Wood.)

noticing its asymmetry), but is representative of those in the British Museum of Natural History. Figure 3–20 is reminiscent of the game in the weekend press in which one is asked to spot the differences between two versions of a cartoon. Scales in butterflies have an important function, to increase lift during gliding. Therefore the orientation of the scales on the

Fig. 3–22 Asymmetrical pattern on a frog, *Dendrobates tinctorius*.

Fig. 3–21 Asymmetrical pattern on a salamander.

wings is important. However, their pattern could be made asymmetrical without affecting gliding performance.

There is a mutant of the fly *Zaprionus* in which left or right scutellar bristles may be missing. Their progeny tend towards re-establishment of bilateral symmetry even though this involves the loss of a matching pair of bristles on either side.

Pattern asymmetry also occurs in vertebrates, and examples are illustrated for a salamander, *Salamandra* (Fig. 3–21), a frog, *Dendrobates tinctorius* (Fig. 3–22) and the common viper, *Vipera berus* (Fig. 2–1b). It is likely that such asymmetry helps to break up the outline of the animal. Such asymmetrical patterns may often be observed on the skin of the adult frog, *Rana temporaria*. The stripes on mackerel fish, *Scomber scombrus*, also show asymmetry (Dr S. E. Reynolds, personal communication).

A remarkable example of pattern asymmetry is seen in the post-larval stage of an eel, *Leptocephalus diptychus*. This has conspicuous spots, four on one side and three on the other. They alternate in position so that all seven spots are seen from either side, since the body is transparent.

4 Asymmetry of Skeletal Fibres

One aspect of the study of differentiation concerns the way in which macromolecular polymers become specifically oriented in extracellular materials such as basement membranes and skeletons. In animals with overall bilateral symmetry, such fibre systems often show asymmetry with respect to the body, and these form the subject of this chapter.

4.1 Vertebrate eyes

An asymmetrical fibre system was discovered in the stromal region of the cornea of chicken eyes by Coulombre and Coulombre (1961). The fibres, which are of the protein collagen, are organized in pairs of layers. In each pair the fibres run parallel in each layer and the two layers are oriented with a 90 degree change in angle. Each of these successive orthogonal pairs is set at an angle to the previous pair, giving a progressive angular displacement (Fig. 4–1). The direction of angular rotation is clockwise and is the same in both eyes of an individual, so that the fibre system is asymmetrical with respect to the body symmetry. The stroma is about 200 μm thick when mature, with individual layers 2 to 4 μm thick. The total angular shift in each eye is about 200 degrees, the angular increment being greater in the last layers to be added. Displacement only appears in layers added after the sixth day from hatching. It has been suggested that the angular displacement serves a mechanical function, giving equal resistance in various directions to pressure from within the eye so as to maintain corneal shape.

The cornea (which is extracellular) is secreted by the corneal epithelium and the angular displacement occurs in a narrow zone next to the cells. Trelstad and Coulombre (1971) have suggested that because of the asymmetry with respect to the whole body, the rotation occurs by the

Epithelium

Fig. 4–1 Diagram of the cornea of a chicken eye showing the 200 degree clockwise shift of the collagen orientations. The sense of twist is the same in both eyes. (After TRELSTAD (1972). In *Comparative Molecular Biology of Extracellular Matrices* (ed. by H. Slavkin), Academic Press.)

collagen molecules self-assembling in a matrix of polysaccharides (keratosulphate, chondroitin-4-sulphate and chondroitin). I have put forward similar arguments for self-assembly of chitin fibres in insect cuticles (see section 4.3).

This type of layer rotation appears to be found in the eyes of all vertebrates except the mammals. Electron micrographs show the rotation of fibres in sections of tadpole eyes. Curiously, the cuticle of arthropod eyes also shows asymmetrical fibre systems (NEVILLE, 1975).

4.2 Geodesic spirals and their functions

The type of fibre asymmetry described for vertebrate eyes in the last section is really a specific example of a more general type of asymmetry. The fibres in the basement membrane, on which sits a layer of cells, show asymmetry with respect to the body as a whole in a wide range of phyla (Table 1). In many cases the fibres (of collagen) are arranged as *geodetics*. A geodetic is the shortest line on a surface joining two points on that

Table 1 Examples of fibre systems which run asymmetrically with respect to the body symmetry.

Phylum	Example
Coelenterata	*Metridium*
Platyhelminthes	Turbellarians
Nematoda	*Ascaris*
	Trichuris suis
Nemertea	*Lineus*
Annelida	*Lumbricus*
Arthropoda	Cuticular chitin
Mollusca	Squid mantle
Chordata	Selachian embryo
Chordata	Teleosts
Chordata	Amphibia
Chordata	Birds, cornea

surface. Individual fibres are often longer than the body. The collagen fibres of the epidermal basement membrane of a tadpole, *Bombinator*, give a false impression of bilateral symmetry when two layers of fibres are superimposed (Fig. 4-2c). However, when they are drawn separately (Fig. 4-2a, b) the true asymmetry is revealed. There are no free ends in the system. In fish, migrating mesoderm cells settle at the intersection points in a similar collagen fibre array. These cells form the scales, whose arrangement is thus determined by the collagen system.

In the Turbellaria and Nemertea such asymmetrically oriented fibres permit great extensibility of the body. The thick epidermal basement membrane contains *inextensible* fibres running around the body in left

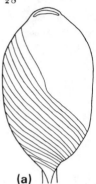 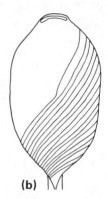 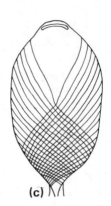

Fig. 4–2 Diagrams of collagen fibre orientations in the ventral epidermal basement membrane of a tadpole, *Bombinator*. Each layer is asymmetrical with respect to the whole body (a) and (b) but superposition of two layers gives the impression of pseudosymmetry (c). (After ROSIN (1946), *Rev. suisse Zool.*, **53**, 133–201.)

and right-handed helices in alternate layers. They work like lazy tongs or trellis fencing, allowing extensive changes in body shape. This requires that no slipping occurs at points where fibres cross, and it also demands an elastic matrix. When the circular muscles of the body wall contract to elongate the body, the fibres become reoriented closer to the longitudinal axis (Fig. 4–3a), whereas when the longitudinal muscles contract to shorten the body the fibres become oriented at a greater angle to the body axis (Fig. 4–3b).

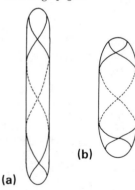

Fig. 4–3 Change in angle of collagen fibres with shape of body.

4.3 Helicoidal structures

There has recently been considerable interest in helicoidal systems, since their interpretation in crab cuticle amd confirmation in insect cuticle. A *helicoidal system* is one in which the component building units are arranged in laminar planes, so as to lie in parallel in any one plane. There is a progressive change in direction from layer to layer, always in one particular sense (Fig. 4–4). Oblique sections of such systems give rise to parabolic patterns (Figs. 4–5 and 4–6a), the components of which are seen particularly clearly in crab cuticle (Fig. 4–6b).

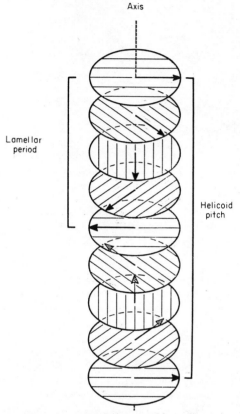

Axis

Lamellar period

Helicoid pitch

Fig. 4–4 Diagram illustrating helicoidal structure. Circles (viewed from above) of planes of construction units orientated in parallel, are drawn for every 45 degrees of rotation about the axis running through the centres of all the circles. Orientation of units is indicated by arrows. (After NEVILLE (1970), *Symp. R. ent. Soc. Lond.*, 5, 17–39.)

In insect cuticle, which forms the insect exoskeleton, there is a helicoidal array of crystallites of the polysaccharide chitin, each of which is 2.8 nm in diameter. They are embedded in a protein matrix to form a composite mechanical material which functions like fibreglass. It has been shown that the sense of twist of insect helicoids is left handed (as in a left-handed corkscrew) on both sides of the body, so that the whole exoskeleton is asymmetrically constructed at this level. (It was incidentally this discovery which prompted the author to write this book.) It is

Fig. 4–5 Diagram showing how parabolic patterns arise in oblique sections of a helicoid. A truncated pyramid of layers is viewed from above. The construction units are arranged parallel to each other forming layers, and their direction changes progressively, and in one sense of twist, from layer to layer. An anti-clockwise helicoid is shown. (After BOULIGAND (1965), *C. R. Acad. Sci.*, Paris, **261,** 3665–8.)

possible to determine the sense of rotation of a helicoid by relating the direction of parabolic patterning to a known plane of sectioning (Fig. 4–7). A section cut so as to include parts of both sides of the body reveals no change in direction of patterning, thus confirming the asymmetry of the system (NEVILLE, 1970).

The structure of arthropod cuticle is analogous to that of cholesteric *liquid crystals*, so that they share similar optical properties. A liquid crystal is a state of matter between those of the solid and liquid states. It possesses order (like a solid) and yet is mobile (like a liquid). One type of liquid crystal is the cholesteric type, found in esters of cholesterol, in certain synthetic polypeptides mixed with various organic liquids, and in DNA mixed with water. (In the latter example, the well-known double helix forms the building unit of the helicoid.)

If the pitch (i.e. the distance taken to rotate 360 degrees through a sequence of layers) is correct, helicoidal systems can produce interference colours according to the formula

$$\lambda = PN$$

where λ is the wavelength of colour reflected; P is the pitch of the helicoid; and N is the refractive index along a light path normal to the component layers of the helicoid. Such a system has the peculiar property of reflecting circularly polarized light. Light entering such a system may be represented as two planes of plane polarized light (defined in Chapter 1)

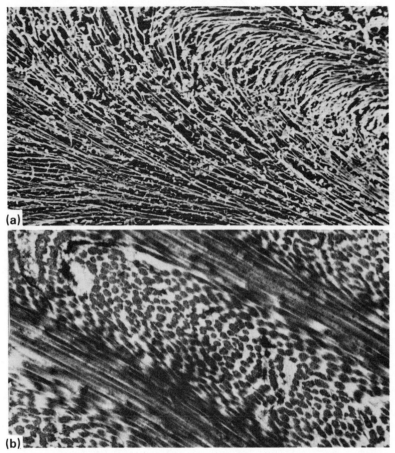

Fig. 4–6 (a) Scanning electron micrograph of fractured surface from Tunicate test (*Halocynthia papillosa*). (From GUBB, D. C. (1975), *Tissue and Cell*, **7**, 19–32.) × 600. (b) Electron micrograph of oblique section through cuticle of a crab (*Carcinus maenas*). (Micrograph by Mrs B. M. Luke, from NEVILLE, A. C. (1970), *Symp. R. ent. Soc. Lond.*, **5**, 17–39). × 35 000.

which are arranged mutually at right angles and with a phase difference of 90 degrees (Fig. 4–8). The resultant of such a wave combination in space is represented by a spiral. Half of the light spirals the same way as the sense of twist of the layers of the helicoid, and this will be transmitted. The other half is reflected and spirals in the opposite direction. For a left-handed helicoid, left circularly polarized light is reflected. Certain brightly coloured beetles in the scarab family reflect left circularly polarized light. (A British example is the metallic green rose chafer, *Cetonia aurata* and another is the prothorax of the common cockchafer, *Melolontha melolontha*; the phenomenon can be seen in the latter by wetting the hairs on the prothorax.) To detect circularly polarized light the beetle is viewed

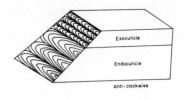

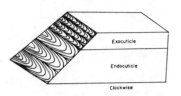

Fig. 4–7 Diagrams to show the method for deducing sense of rotation of a helicoid from the direction of parabolic pattern on a known oblique face. (From NEVILLE and LUKE (1971). *J. Insect Physiol.*, **17**, 519–26.)

through a piece of plastic sheet (whose molecular chains are strongly oriented in one direction, and which is cut to a thickness appropriate to retard light in one plane by 90 degrees relative to a plane at right angles to it. This converts the circularly polarized light coming from the beetle into plane polarized light. If we now place a piece of polarizing filter over the

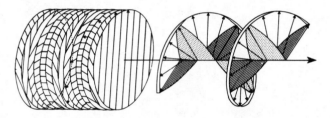

Fig. 4–8 Diagram relating the sense of rotation of planes of constructional units in a helicoid (anti-clockwise looking away from the observer) to the sense of rotation of transmitted circularly polarized light (right approaching observer). Reflected circularly polarized light would have the opposite sense of rotation (i.e. left). (From NEVILLE and LUKE (1971),*J. Insect Physiol.*, **17**, 519–26.)

plastic sheet, it will cut out the light from the beetle when rotated so as to cross the plane of polarized light emerging from the plastic. The fact that the whole beetle appears black, and not just half of it, shows that the sense of the helicoid is constant (and hence asymmetrical) over the whole exoskeleton. Using this quick test we have found that all of the several hundred species of scarab beetles (in the British Museum of Natural History) which show these properties (optical activity), reflect left circularly polarized light. Therefore they have left-handed helicoidal structure throughout the skeleton. This asymmetry led me to propose that arthropod exoskeletons are formed by self-assembly of a cholesteric liquid crystalline stage which is later stabilized by crystallization.

The asymmetrical helicoid architecture in arthropod cuticle has an effect on two related systems. A set of tubes, the pore canals, run through the thickness of the cuticle from the underlying layer of epidermal cells. They are twisted into anti-clockwise ribbons by the surrounding helicoidal chitin architecture (Fig. 4–9).

In the beetle *Plusiotis resplendens*, which is a brilliant metallic golden colour, the cuticle contains up to 70% uric acid by volume. This is

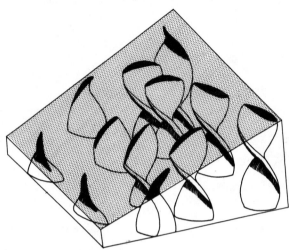

Fig. 4–9 Diagram showing the origin of a parabolic pattern (in black) of pore canals, in arthropod cuticle. The pattern arises on an oblique plane of section (stippled) cut through a group of canals which twist in unison. (From NEVILLE and BERG (1971), *Palaeontology*, **14**, 201–5.)

incorporated as small needle-shaped crystals whose orientation faithfully follows that of the surrounding helicoidal chitin system.

Whereas the pitch of arthropod cuticle helicoids ranges between 0.2 to 30 μm, that of the tunicate *Halocynthia papillosa* approaches 100 μm. The test is constructed from a helicoidal array of tunicin (animal cellulose) fibres, whose diameter is fortunately large enough to be resolved in the scanning electron microscope. Figure 4–6b visualizes a fractured surface from such a test and shows the parabolic pattern made up of differently oriented fibres. As with the arthropod cuticle, the helicoid is asymmetrical with respect to the whole individual, being left-handed throughout the test. The helicoid of a nematode eggshell, *Trichuris suis*, is also asymmetrical with respect to the whole shell (Wharton, personal communication).

Since helicoidal structure has been found in all of the arthropod exoskeletons so far examined, it follows that *helicoids are the commonest form of animal extracellular architecture*, both in numbers of individuals and in numbers of species. Apart from the arthropoda, helicoids (as recognized by parabolic patterning) are found in the extracellular structures of many other phyla (Table 2). A general review of helicoids is given by BOULIGAND (1972). It would be interesting firstly to confirm the helicoidal nature

Table 2   Helicoidal structures in organisms, as indicated by parabolic patterns.

Phylum	Example
Protozoa	Dinoflagellates (Chromosomal DNA)
Coelenterata	*Aurelia* (podocyst cuticle)
Platyhelminthes	*Discocelides* (Subepidermal membrane)
Annelida	*Diopatra* (oocyte envelope)
Rotifera	*Philodena* (integumental shield)
Nematoda	*Heterodera* (cuticle)
Arthropoda	Crustacea, Insecta, Arachnida, Myriapoda (cuticle) *Hyalophora* (egg shell) *Sphodromantis* (oothecal protein)
Mollusca	*Buccinum* (periostracum)
Echinodermata	*Thyone* (body wall)
Chordata	*Cynolebias* (oocyte shell) *Halocynthia* (tunicin test)

suggested by these patternings, and secondly to discover whether they are all asymmetrical with respect to the whole organism. So far we know that the arthropod cuticle and a tunicate test are left-handed helicoids. Are all the others? And if so, is there a fundamental common chemical denominator causing this?

The cuticle deposited after ecdysis (endocuticle) in some adult insects (beetles, bugs, stick insects and dragon flies) superficially resembles plywood in the orientation of layers. Electron microscopy has, however, revealed a more complex arrangement, since a change from one layer of undirectionally oriented microfibrils to another layer is not abrupt. Instead there is an intervening region with helicoidal orientation. Figure 4–10 illustrates this. The orientation of the first endocuticle layer deposited preserves the bilateral symmetry in beetles. However, asymmetry appears in the second and subsequent layers (Fig. 4–11). The

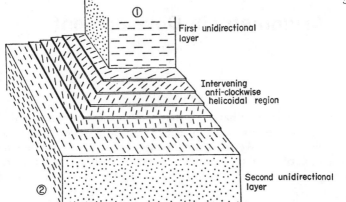

Fig. 4-10 Diagram of chitin fibre orientations in adult beetle cuticle. Sandwiched between two layers in each of which the fibres run in parallel, is a region of helicoidal structure. (From ZELAZNY and NEVILLE (1972), *J. Insect Physiol.*, **18**, 2095-121.)

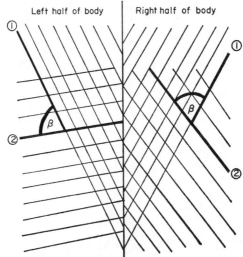

Fig. 4-11 Diagram to illustrate the loss of bilateral symmetry in the pattern of chitin fibre orientation between the first and second layers of endocuticle in an adult beetle. Layer (1) is orientated symmetrically about the midline of the insect, which divides the diagram in two halves. An anti-clockwise helicoidal rotation through β degrees (in this case 70 degrees) then occurs on both sides of the body so that layer (2) becomes asymmetrical. (From ZELAZNY and NEVILLE (1972), *J. Insect Physiol.*, **18**, 2095-121.)

orientations are controlled to an accuracy of ± 3.3 degrees by the cells which secrete the layers, and experiments have shown that the mechanism for this involves a polarity gradient of information between cells (see Chapter 5).

5   Asymmetry in Development

5.1   Spiral cleavage in gastropods

We have seen in Chapter 3 that most species of gastropods have dextrally coiled shells. Some varieties are sinistral while a few species are normally sinistral. We can trace the origin of this back to the plane of orientation of the mitotic spindles when the fertilized egg is undergoing cleavage from the two cell to the four cell stage (Fig. 5–1). The result of this

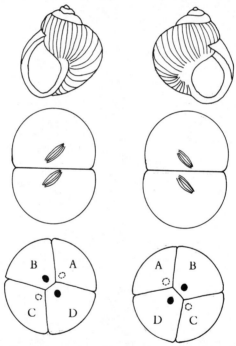

Fig. 5–1   The effect of sense of spiral cleavage on snail twisting. Sinistrals on the left; dextrals on the right. The oblique cleavage plane of the mitotic spindles in the two cell stage gives rise to oblique furrows in the four cell stage. The latter pair are related as mirror images, as are the resulting snails. (After HUXLEY and DE BEER (1963), *The Elements of Experimental Embryology*. Hafner, New York.)

is to establish one of two arrangements (which are related enantiomorphically), depending upon the plane of the spindles (Fig. 5–1). The arrangement of the four cells is reminiscent of that of the atoms on an asymmetric carbon atom (Fig. 1–1). There are two basic types of cleavage in animals. One is *radial cleavage* shown in Fig. 5–2, which leads

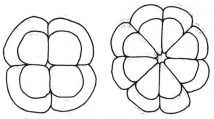

Fig. 5-2 Diagram of radial cleavage of an embryo from the eight to the sixteen cell stage.

to bilateral symmetry. This is contrasted with *spiral cleavage* (Fig. 5–3) in which the successive cell divisions result in alternate displacement, clockwise alternating with anti-clockwise. Spiral cleavage establishes asymmetry, and is found in Polyclad Turbellaria (Platyhelminthes), Nemertini, Annelida, Sipunculida and Mollusca (with the exception of the Cephalopoda which are bilaterally symmetrical). A clockwise third cell division leads to a dextral gastropod; an anti-clockwise one to a sinistral example. By the time cell division has proceeded to the stage

Fig. 5-3 Diagram of spiral cleavage of an embryo from the eight to the sixteen cell stage. The third division here is clockwise, the fourth anti-clockwise (arrows).

shown in Fig. 5–4, a prominent cell (4d) can be seen. This gives rise to the right mesoderm band which forms the retractor muscle. We have seen already (section 3.15) the importance of this muscle in torsion. The cell 4d

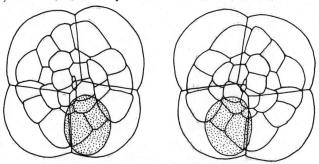

Fig. 5-4 Cleavage asymmetry in Molluscs. The position (dotted) of the large mesoderm cell (4d) is reversed in sinistral (on left) or dextral (on right) cleavage. This cell produces the muscle responsible for torsional asymmetry. (After MORGAN (1927), *Experimental Embryology*. Columbia University Press.)

occupies a mirror-imaged position in dextral as compared with sinistral individuals.

5.2   Determination of asymmetry in snail eggs

The self-fertilized offspring from a hermaphrodite individual of the pond snail, *Limnaea peregra*, were either all dextral or all sinistral, irrespective of the parent's sense of twisting. Sturtevant (1923) therefore proposed that direction of coiling is in the main determined by a pair of Mendelian factors, with dextral dominant and sinistral recessive. The inheritance of sinistrality in *Limnaea peregra* was extensively investigated (6000 broods and one million snails) by BOYCOTT, *et al.* (1931). They concluded that sense of twist is determined not by the chromosomes of an individual but by those of its parents. This is a classical example of maternal or delayed inheritance. The twist is already determined in the unfertilized egg and not by the sperm which fertilizes it. The effect of the sperm chromosomes on sense of twist is not seen until the next generation. The system is Mendelian but with a generation delay, so that segregation is by broods in the third generation rather than by individuals in the second generation. Dextrality is dominant: but there is a tendency for some dextrals to appear even when sinistrals are expected. One of the sinistral lines in these experiments gave only 8 dextrals out of over 20 000 snails. That the results were not due to abnormal behaviour of chromosomes was shown by the normal Mendelian inheritance of albino character in the same snails. Dextral and sinistral *Limnaea peregra* can in fact interbreed. There is no evidence of difference in mortality between dextrals and sinistrals. Previous workers experienced difficulty in crossing sinistral and dextral *Helix pomatia*.

5.3   Insect gynandromorphs

Sex changes in insects cannot be brought about by removal of gonads, as the sex is determined in each and every cell by chromosomes. It thus becomes possible to have insect freaks in which part of the body is one sex and the rest is the other sex. In butterflies the males have XX sex chromosomes. If one of these is absent the sex is female. Thus, if one of the X chromosomes is lost during the first nuclear division of the fertilized zygote, that cell becomes female while the other remains male. Cleavage being determinate in insects, one of these two cells gives rise to the left half of the body and the other to the right half. This leads to an individual with half the body male and the other half female (a *bilateral gynandromorph*). In species with marked sexual dimorphism this gives striking results. For example, in the common blue butterfly (*Polyommatus icarus*), the male has blue wings whereas the female has brown wings. A bilateral

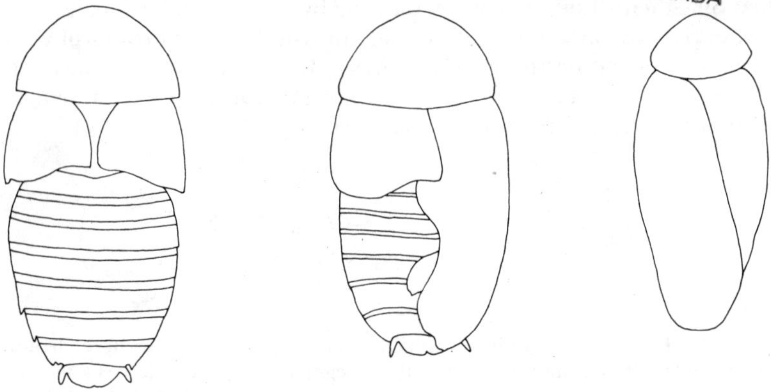

Fig. 5–5 Female (left) with reduced wings and male (right) with full-sized wings of a cockroach, *Byrostria fumigata*. A bilateral gynandromorph (centre) has a female left half with reduced wings and a male right half with full-sized wings. The appendages are not shown. (After WILLIS and ROTH (1959), *Ann. ent. Soc. Amer.*, **52**, 420–9.)

gynandromorph thus shows outstanding asymmetry, having two blue male wings on one side of the body and two brown female wings on the other side. Coloured illustrations of such examples in butterflies are shown in Ford (1945). Gynandromorphy may occur more frequently in some districts than in others. Ford records them as occurring more often in Irish common blues than in English examples. In the silver-washed fritillary butterfly (*Argynnis paphia*) there is a dark variety found only in females, and mainly restricted to the New Forest. Bilateral gynandromorphs are known in which half is a normal male and half female variety *valezina*. Even more bizarre are the two moths bred by Whicher (1915), obtained by crossing a male eyed hawk moth (*Smerinthus ocellatus*) with a female poplar hawk moth (*Laothoe populi*). They were male eyed hawk on one half and hybrid (poplar × eyed) on the other half.

Gynandromorphs also occur in larvae. Cockayne (1935) crossed two commercial silk moths (*Bombyx mori*) hatched from a female larva with a zebra pattern and a male larva which was white. Two of the resulting larvae had a zebra pattern on one half and were white on the other. A larva has also been bred with a black male right-hand side and a white female left-hand side.

As well as patterns, whole body form may be involved. Thus, males of the cockroach *Byrostria fumigata* are winged whereas females are wingless. A remarkable bilateral gynandromorph described and illustrated by Willis and Roth (1959) has wings on the male half and none on the female half (Fig. 5–5).

If a male chromosome is lost at a later stage of nuclear division than the first one, then all subsequent cells arising by mitosis will be female. This gives rise to asymmetrical sexual mosaics. Gynandromorphy is commoner in some species than others. It is particularly frequent in *Bombyx mori*. An example of a gynandromorph involving just the antennae is shown in Fig. 5–6, for *Habrobracon juglandis*.

Fig. 5–6   Head of a gynandromorph *Habrobracon juglandis* seen from the front. The right side has a male antenna with 19 segments; the left side has a female antenna with 12 segments. (After CLARK, PETERS and BRYANT (1973), *Devel. Biol.*, 32, 432–45.)

In a paper which offers exciting possibilities for further work, Ikeda and Kaplan (1970) have made use of genetic mosaics in *Drosophila* to locate the sources of motor neuron output. Using a mutant (hyperkinetic) which shows leg shaking during ether anaesthetization, and which is found only in female homozygotes and male hemizygotes, they bred several mosaic gynandromorphs. The shaking response in bilateral examples is restricted to the male side, as are the relevant bursts of motor nerve output. Even when the whole abdomen or the whole head in mosaics was female (as judged by external form), typical male leg shaking occurred only on the male half of the thorax. They conclude that there is correspondence between the genotype for thoracic cuticle and for the motor regions of the ganglia responsible for these electical outputs. This is consistent with the common origin during development of neuroblasts (forming neurons) and of the precursor cells of adult epidermis.

5.4   Labiopedia in insects

Development occasionally goes wrong in insects, and one error which occurs is that the labium of the mouthparts becomes replaced by an extra pair of legs. Daly and Sokoloff (1965) illustrate such an example for *Tribolium confusum*, the individual concerned being thus an 8-legged beetle! The extra pair of legs are quite asymmetrical.

5.5   Asymmetry in bed bugs

In male bed bugs (*Cimex*) the genitalia are asymmetrical. There is slight asymmetry in the last instar larva, the left clasper being enlarged. Marked asymmetry appears at the last ecdysis and is said to be caused by asymmetry in the symbiotic organ. An intermediate stage can be experimentally produced by juvenile hormone treatment, achieved by parabiosis with a third instar larva (or presumably by appropriate injections of synthetic juvenile hormone analogues now commercially available). This produces a left clasper intermediate in size between that of a final larva and an adult. The asymmetry here is therefore not an all or nothing phenomenon, but is a graded response.

5.6   Cell polarity gradients

A concept which is widely used to explain developmental co-ordination is that of the cell polarity gradient. This is an important, and as yet unsolved, problem in developmental biology. A row of cells are assumed to be related by a stable gradient of information, by reference to which they can determine their precise position, even if moved into the wrong place experimentally. It is not known whether the gradient is chemical or physical. By using the gradient a cell is able to know if it is in the correct place, and also able to control the specific direction of any polarized extra-cellular products which it secretes. One clear case where this has been demonstrated is that of chitin fibre direction control (see Chapter 4) in adult beetle cuticle.

The standardized experiment for investigating gradient effects is to cut out a square of tissue, rotate it through a known angle, replace the graft,

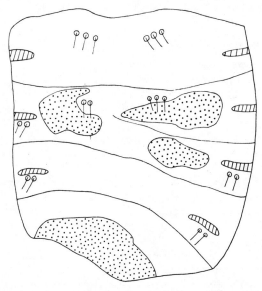

Fig. 5–7   Abdomen of *Oncopeltus* with a cell gradient defect, showing asymmetry of bristles, pigmentation (dotted), sclerite shape, and muscle insertions (hatched). (After LAWRENCE (1970), *Adv. Insect Physiol.*, **7**, 197–266.)

and then (in the case of an insect) to allow the animal to continue its moulting cycles. The results are then analysed in a later instar. When cells are displaced in this way from their normal level in a gradient, they eventually adapt to their new environment, but this process takes time. The cells at the centre of a rotated graft retain their original polarity for longer than those at the edges (which are in contact with those of clashing gradient). In experiments by Caveney, grafts were performed on larvae of the mealworm beetle (*Tenebrio molitor*), and the results analysed in the subsequent adults. We have seen in Chapter 4 that the pattern of chitin fibres in the first endocuticle layer of adult beetles is bilaterally symmetrical, making it suitable for analysis of gradient perturbation experiments. When a square of integument (cuticle plus the cells which secrete it) was rotated through 90 degrees, subsequent deposits behaved exactly as predicted if a cell polarity gradient controlled fibre orientation. The asymmetrical endocuticle layers laid down after the first one rotate in orientation as in other beetles, with a specific and accurate sequence of direction. These directions are maintained relative to layer 1 in endocuticle deposited in both 30 degrees and 90 degrees rotated graft areas, even though each layer is now orientated at the wrong angle with respect to the whole insect.

When gradient control breaks down, many features may become asymmetrical. Figure 5–7 shows asymmetry of bristles, pigmentation, sclerite shape and muscle insertions in a defective bug, *Oncopeltus*.

5.7 Protochordate asymmetry

In *Branchiostoma lanceolatum* (the Amphioxus made famous by embryologists) the larva is markedly asymmetrical with gill slits appearing first on the left-hand side, having migrated from the right where they were formed. The mouth begins on the left and migrates to the centre, but receives nerves and blood supply from the left only. The young larva, swimming by epidermal flagella, rotates clockwise as seen from the front, so that the mouth on the left is always in front. A related genus is even called *Asymmetron*!

On the gill slits of *Saccoglossus*, *Ciona*, *Ascidia* and *Branchiostoma*, the cilia beat with anti-clockwise metachronal waves on both sides of the animal. This is interesting because the gill slits are paired structures. Knight-Jones and Millar (1949) suggest that this may be caused by the telotroch belt of large cilia present in the larvae. These large cilia beat anti-clockwise and since they lie next to the developing gill slits at metamorphosis, their currents could well influence the ciliary beat on the gills slits.

In the colonial sea squirt, *Botryllus schlosseri*, asymmetry involves the gut, larger testes and more ova lying on the left, and the heart and reproductive buds on the right. Sabbadin (1956) experimentally reversed this asymmetry (*situs inversus*) in buds which were stimulated to resume development after remaining dormant for some time, during which the parent zooid degenerated. The parental zooid determines the asymmetry of a bud during its development. An inverted zooid passes on its inversion to asexually produced buds. A normal zooid produces normal buds. But inversion is *not* passed on to sexually reproduced offspring; it does not therefore involve the genotype. Izzard showed that the young buds lose their determined sense of asymmetry during the degeneration of the parent. It would be of great interest to know how the information to become inverted is passed from parent to bud.

5.8   Gut asymmetry in Amphibia and fish monsters

Experimental embryologists have shown that *situs inversus* of gut and heart in newts and frogs is determined by a factor in the gut roof. At the developmental stage when the neural folds are still open, a square piece of presumptive neural tissue plus underlying gut roof endoderm is rotated through 180 degrees and replaced. The individual then grows with *situs inversus*, i.e. with the stomach on the right, intestine and heart twisted opposite to the normal direction, and the spiracle on the wrong side (Fig. 5–8). The gut roof is the determining factor, since reversal does not occur if only neural cells are rotated. It is tempting to suggest that a gradient mechanism is operative here (see section 5.6). However, more

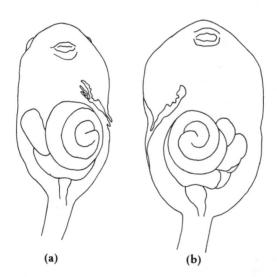

(a)   **(b)**

Fig. 5–8   Experiment on gut reversal in a tadpole, *Bombinator*. (a) Ventral view of normal embryo. (b) Embryo produced by turning a square of dorsal surface and gut roof through 180 degrees in a neurula. This has resulted in gut reversal and the spiracle on the wrong side. (After MEYER (1913), *Arch. Entwmech.*, **37**, 85–107.)

recent experiments showed that reversal still occurs even when the square of tissue is replaced without rotation.

When a developing blastula of *Triton* was partially constricted in the plane of bilateral symmetry by means of a fine hair, a double-headed monster was produced. The left member was normal, but the right member showed *situs inversus*. Such double-headed monsters occur naturally in trout. When joined between stomach and cloaca the left member is normal and the right member shows *situs inversus* (like human Siamese twins they are enantiomorphic book-ends, see section 3.14). If, however, the join is further back, there is normal gut coiling in both members.

If they are separated at the two cell stage, newts do not show any more *situs inversus* individuals than the 3% found normally. Therefore the *asymmetry bias factor* is not yet present at the two cell stage.

Besides the axial polarity gradient system which determines the anterior and posterior of the body, HUXLEY and DE BEER (1963) postulate a further gradient running across the body from left to right which is concerned with asymmetry. Its establishment is also connected with the dorso-ventral gradient which determines bilateral symmetry. Both gradients intensify as cleavage proceeds, perhaps because more cells join the sequence.

Selected aspects of developmental asymmetry are reviewed by OPPENHEIMER (1974).

6 Asymmetry in Behaviour

6.1 Asymmetry in arthropod behaviour

Asymmetry of crab walking is well known from the popular phrase 'sideways like a crab'. It is performed by synchronous elevation of the trailing legs with flexure and closing of their distal segments. Elevation, extension and closing of the legs on the leading side occurs. There is a central nervous programme with sensory modification.

In the crab *Cancer*, the right maxillipeds beat during walking to the left and vice versa. Stationary crabs show bias in maxilliped beating, to one side or the other. If a crab with the left ones active is stimulated from the right-hand side, it switches to beating the right ones, indicating a preparation at the central level to walk away from the stimulus to the left. Burrows and Willows (1969) suggest a model involving left versus right reciprocal inhibition of neurons supplying the muscles involved.

In the hermit crab, *Pagurus granosimanus*, the sensory units on the left side of the asymmetrical abdomen are more sensitive to mechanical stimuli than those on the right-hand side. This is correlated with the right side lying next to the host shell and the left side facing away from it. Hence the left side needs to convey more information, e.g. about location of commensals and parasites which also inhabit the shell. The third ganglionic roots are exclusively motor, with more spontaneous motor output on the left side. There is a loss of motor fibres in the first ganglionic roots, correlated with the absence of pleopods on the right side.

Locusts (*Schistocerca gregaria*) flying in the dark in front of a wind tunnel, show asymmetrical motor nerve output to flight muscles on opposite sides of the body (WILSON, 1968). When suitably suspended, some locusts always roll clockwise and some anti-clockwise, about the long axis of the body. Figure 6–1 shows electrical recordings from the right and left subalar muscles of the hind wings. At the point marked 'ON', the light was switched on, causing compensatory motor output induced by the sense organs to counteract rolling. Extra force is generated on the side requiring it: this is done by double firing of the muscle, i.e. firing it a second time after each normal firing.

Further asymmetry in insect flight systems is shown by the unequal distribution of indirect flight muscle fibres in adult flies, e.g. *Aedes aegypti*, *Syrphus viridiceps* and *Anisopus fenestralis*. As in locusts, this asymmetry must also be compensated during flight by sensory reflexes. Similar sensory compensation prevents milkweed bugs (*Oncopeltus*) from walking in circles, as there is an asymmetrical nervous output to the hind legs in

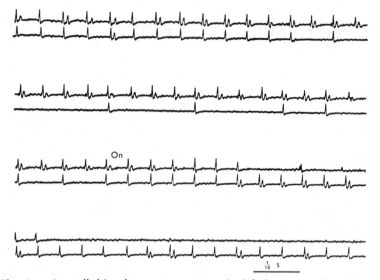

Fig. 6–1 Laterally biased motor nerve output in right (upper trace in each pair) versus left (lower trace) metathoracic subalar muscle in a locust flying in a wind tunnel. Turning on the light (On) produced a strong compensatory reaction in the less active side. (From WILSON (1968), *J. exp. Biol.*, **48**, 631–41.)

unstimulated preparations. The central nervous system causes bias in turning left or right in an individual bug. This bias is corrected in a walking bug by an optomotor reaction. The turning bias in one bug could be constant for several days on a Y-maze globe apparatus. In this type of experiment the insect is held fixed. It walks on a cut out globe held by its legs, and which turns as walking movements occur. The globe is cut away to leave walking tracks, which diverge into left and right tracks at four points around the globe. The insect therefore has free choice of turning without visual feedback since the body does not move.

The moth *Automeris* shows a rocking behaviour, consisting of side to side oscillations of the head. If rocking is initiated a light signal to one eye causes the rocking to be asymmetrically biased to that side. Individual grasshoppers show asymmetrical behaviour as they preferentially use one foreleg for grooming. Aidley and White (1968) demonstrated asymmetrical motor output to the sound-producing muscles of a Brazilian cicada, *Fidicina rana*. The left and right muscles involved are fired exactly out of phase so as to double the frequency.

6.2 Asymmetry in mollusc behaviour

The scallop, *Pecten*, swims horizontally with the right-hand side upwards. Though both balancing organs (statocysts) are present, the right one is vestigial and all orientation and control of swimming reflexes are initiated exclusively from the left statocyst. In *Octopus*, the statocysts on the two sides of the body are orientated mutually at right angles in the horizontal plane, so that in the restrained animal the input from both is additive.

6.3　Ear asymmetry in owls

In Tengmalm's Owl (*Aegolius funereus*) the skull is asymmetrical to give different vertical positions to the left and right ear apertures (Fig. 6–2).

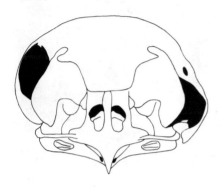

Fig. 6–2 Skull of Tengmalm's owl, *Aegolius funereus* seen from the front, to show the asymmetry of the ear apertures. (After NORBERG (1968), *Arkiv. Zool.*, **20**, 181–204.)

The latter are covered over with sound-transparent feathers of the facial disc. On intact specimens the asymmetry may only be seen by pulling these feathers forward. The external ear asymmetry allows judgement of direction of high frequency sounds in the vertical plane, without the need to tilt the head. In an owl with symmetrical ear apertures, sound direction can be judged in the horizontal plane without tilting the head, but tilting is required for estimates in the vertical plane. In *Aegolius* the right ear has its maximum sensitivity in an obliquely upward direction, the left ear in an obliquely downward direction. The asymmetry can be useful for sound frequencies between 10 000 and 16 000 cycles per second, but not below 10 000 c/s (NORBERG, 1968).

Young owls learn to interpret the asymmetrical information from both ears by tilting their heads alternately to left and right in the vertical plane, often through more than a right angle (Fig. 6–3). When low frequencies sound equally loud in both ears, but higher frequencies are louder in the right ear, the young owl learns that the sound is coming from above and not from the right. The owl can then use this ability to pinpoint the direction of the sounds of potential prey rustling in the vegetation or in the snow. The mechanism is so good that several owls (*Tyto alba pratincola*, *Strix varia* and *Asio otus wilsonianus*) can locate their prey by sounds in total darkness. In *Tyto alba* the prey can be pinpointed to one degree accuracy in both vertical and horizontal planes. Using microphones implanted

Fig. 6–3 A young Tengmalm's owl (*Aegolius funereus*) learning to locate sounds in the vertical plane by tilting its head. (From NORBERG, Å. (1973), *Svensk Naturvetenskap*, **26**, 89–101.)

in both ears, Payne (1962) showed that the owl would directly face a sound source if it turned its head so as to receive equal sound in both ears.

6.4 Asymmetry in marine mammals

It has long been known that some toothed whales and dolphins have markedly asymmetrical skulls (Fig. 6–4). D'Arcy Thompson (1917) regarded this as resulting from growth during swimming with a spiral

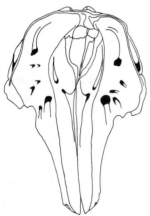

Fig. 6–4 The asymmetrical skull of a Narwhal, *Monodon*, in dorsal view. (After VAN BENEDEN and GERVAIS (1880), *Ostéographie des Cétacés vivants et fossiles*. Paris.)

torque. However, this cannot be so, since asymmetry is already present in the skull of the embryo. The single nostril (blowhole) is asymmetrical, usually lying to the left of the midline. The right-hand muscle and sacs of the blowhole are larger. In view of the previous section on owl skull asymmetry, it seems reasonable to suggest that whale skull asymmetry may also be an adaptation for sound communication in more than one plane. The sound producing abilities of these animals are well known.

In the Narwhal it is usual for only one tooth to mature. This tusk may grow to nine feet and carries a screw along its length. In rare examples, two tusks may grow, both of which carry a right-handed thread.

6.5   Asymmetry in human brains

There are really two brains in a mammal, present as two halves divided by the midline of the body. The right half of the brain controls the left-hand side of the body and vice versa. If part of one half of the brain is damaged, all is not lost, since the corresponding part of the other side can take over its function.

The two halves of the brain are joined by commissures (which contain millions of nerve fibres). In cats, monkeys and man, cutting these commissures has little obvious effect on behaviour. Some humans in fact lack the largest of these commissures from birth, so that its function was a mystery for a long time. One cynical viewpoint was that it merely serves to spread an epileptic fit from one side of the brain to the other. Sperry (1964) investigated brain functions by split brain experiments, in which these commissures were severed by surgery. This prevents transfer of learned functions from one side of the brain to the other. The independence of the two brains was revealed, since each half of the brain could be taught opposite solutions to a simple problem. For example split brain cats and monkeys could be taught to prefer (because of a subsequent reward) pushing a square button to a round one, by covering one eye. By repeating the training via the other eye, the other half of the brain could be taught to prefer pushing the round button.

In humans there is an asymmetrical division of functions between the two halves of the brain. Right-handed people usually store language and mathematical functions in the left half of the brain. This can be determined from patients who have suffered local brain damage. If the left cerebral hemisphere is damaged in young children, they can develop language capacity in the right hemisphere. Most information on brain asymmetry comes, however, from split-brain preparations. In a right-handed split-brain human, language is stored in the left hemisphere. He cannot write meaningfully with the left hand which is controlled by the right hemisphere. Neither can he obey verbal commands using the left hand or left leg. But in a child with the cerebral commissure missing from birth, language centres were developed in both halves of the brain. However, damage to one hemisphere early in life is overcome better by cats than by monkeys, which in turn recover better than humans. This is because cats and lower mammals use the two halves of the brain symmetrically, so that either side is expendable. In monkeys a certain amount of asymmetrical division of function is appearing and mountain gorillas show a preference for right handedness, since 59 out of 72 used the right hand first in chest-beating displays. Asymmetry of function is an advanced feature of brain evolution. Split-brain humans can in fact perform two tasks as fast as a normal person can do one. If the brain was divided early enough in a child, it is possible that both hemispheres would develop analytical ability normally achieved only by the left hemisphere.

Brain asymmetry appears to be one of the fundamental characteristics of the human race. It is as if the brain has two separate 'selves', each striving for control. A table showing the probable asymmetrical division of functions in a human brain is given below (Table 3). Since right-handedness dominates all human races the table is constructed for a right-handed person. To me it is fascinating that there is almost a division into the sciences and the arts, the famous two cultures of C. P. Snow

Table 3   The asymmetrical division of function in a normal right-handed human brain.

Left hemisphere	Right hemisphere
Analytical thinking	Orientation in space
Logic	Arts and crafts
Mathematics	Face recognition
Verbal functions	Musical pitch recognition
(speaking, writing,	Dancing
(hearing and reading	Sports
words)	Depth perception

(1965). In most people one half of the brain dominates their personality, and one side of the face reveals true character more than the other side.

GAZZANIGA (1967) has pursued the split brain method of investigation in humans. Can the two cerebral hemispheres have separate thoughts and emotions? The left hemisphere can describe a spoon verbally in a split-brain subject: the right hemisphere cannot do so. Yet the right hemisphere can control the left hand to identify the spoon by touch. While the right hemisphere achieves the correct solution to a tactile test, it is at the same time offering the wrong solution verbally.

The language ability of the right hemisphere is very restricted. Thus, some split-brain subjects could mechanically spell the word 'love' when the four letters were presented to the left hand (i.e. right hemisphere); they could not see the letters and knew only that they would potentially spell only one word. Yet although the right hemisphere produced the correct answer, the patients were unable to name the word, since that would require input from the right hand to the language centre in the left hemisphere. The right hemisphere can understand the word 'pencil', but not verbs such as 'frown' and 'smile'. It cannot make plural words. Yet before specialization occurs (up to 4 years old), the right hemisphere is as competent at language functions as the left. The right hemisphere can, however, control the left hand to copy a drawing of a cube. The left hemisphere cannot control the right hand to do this task, even though the right hand is the one normally used for drawing.

In humans it is known that the right half of each eye sends its input to the right cerebral hemisphere. The input from the left half of each eye goes to the left hemisphere. Sperry used this in a further demonstration that the two hemispheres can behave separately and conflictingly. A split-brain patient looks at the word 'HEART'. The apparatus is designed so that 'HE' is seen by the right half of both eyes whereas 'ART' is seen by the left-hand halves. When asked to name the word the subject said 'ART' because that was the message sent to the left hemisphere (the speech centre). When asked to point with the left hand (right hemisphere controlled) to one of two cards containing 'HE' and 'ART' the answer was 'HE'. The two hemispheres gave different answers.

Besides evidence of brain asymmetry from accident cases and split-brain patients, there is a further method of demonstration which can be performed on intact brains. This consists of using an electroencephalograph to record electrical output. During a verbal task the α-rhythm is stronger in the right hemisphere, whereas during a spatial exercise it is stronger in the left hemisphere. Since this rhythm appears in parts of the brain which are turned off, the above result confirms that the left hemisphere acts during speaking and the right one during manipulation. Attempts, at too late an age, to force a left-handed person to become right-handed, can lead to interference with the speech centre, manifest as stuttering.

6.6   Spiral movement in man

Blindfolded subjects walk (unbeknown to themselves) in spirals (SCHAEFFER, 1928). When a person loses his way he walks in circles. Spiral walking can be conveniently investigated when snow is on the ground, as the tracks are then self-recording (Fig. 6–5a and b). To keep the mind occupied during the tests, subjects were asked to fix their minds on some distant destination. The tests must of course be performed on level ground. Spiral progression is also seen when blindfolded subjects swim, row a boat, or drive a car (Fig. 6–5c). The last example shows that rhythmical motor output to the limbs is not required for the spiral effect to be produced. It also shows that the explanations of earlier workers were wrong—it had been suggested that morphological asymmetry was responsible for spiral locomotion. Further, experiments in which the subjects walked first forwards, then backwards, also eliminate this explanation. In this case a stronger right leg walking forwards would effectively be on the left-hand side when the subject turned through 180 degrees and carried on walking in reverse. Yet the sense of spiralling did not alter (Fig. 6–5b). Also, there is no correlation between sense of spiralling and left or right handedness. None of Schaeffer's 34

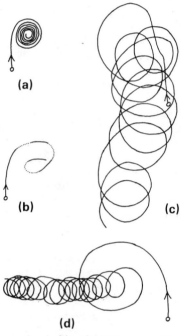

Fig. 6–5 Spiral locomotion in blindfolded humans. Each record begins at the small circle. The subjects tried to head in the direction corresponding to the top of this page. (a) Walking: outer circle diameter of 15 metres. (b) Walking alternately forwards (solid line) and backwards (dotted line) for fifty paces each, turning 180 degrees at each change. (c) Car driving: total distance covered was 5.16 kilometres. (d) Swimming: circle diameter about six metres. (After SCHAEFFER (1928), *J. Morph. Physiol.*, **45**, 293–398.)

blindfolded subjects could walk or swim a straight path for more than a very short distance. Failing a snowfield (or a sandy beach) for recording results, veering to one direction or the other can also be tested in a large hall with the aid of a collaborator. There is some evidence that girls deviate from the straight path more than boys. For most people, a full circle is walked in about half an hour, but the turning circles with the other methods are much tighter. Swimming tracks (Fig. 6–5d) were plotted on tracing paper using a wide angle lens camera with ground glass screen; mounted about 12 metres above the water. All experiments began with a dive and blindfolding was by pulling a swimming cap down over the eyes. Correction was made for perspective. It is not recommended that the blindfold car-driving be attempted on a motorway. (Schaeffer incidentally used an old Ford coupé.)

6.7   Spiral movement in other animals

Bullington (1925) found that 102 species of ciliate Protozoa always spiralled anti-clockwise when they swam, whereas another 62 species spiralled the other way. From this, and observations on flagellate Protozoa and on rotifers, he concluded that species lacking special sense organs (e.g. eyes and statocysts) always rotate in one direction, whereas higher organisms are able to compensate by sense organs and may rotate in either direction.

In the Atlantic palolo worm (*Leodice viridis*) the ciliated larva rotates in an anti-clockwise direction. The turns become closer and closer with age, until at six days old it stops swimming, sinks to the bottom and begins to crawl. When it later swims as a worm, the tail leads, but the direction of twist is still anti-clockwise.

If the eighth cranial nerve is severed and the eye removed on both sides of the body, dogfish, frogs and toadfish all swim with helical tracks.

There is a brain disease in fish which causes them to swim in circles. It is caused by a Myxosporidian Protozoan, one of the commonest of which (*Myxosoma cerebralis*) attacks trout. The disease can often be observed in tropical aquarium fish.

Goslings hatch from the egg by making an anti-clockwise perforation as seen from the outside. This correlates with the position of the head since the neck always coils to one side (P. C. J. Caldwell, personal communication).

6.8   Functional uses of asymmetry

In many cases we can see the usefulness of asymmetry to an animal. It is a means of carrying the normal division of labour in a bilateral animal one step further. For example, heterochely in Crustacea permits the left and right claws to specialize for different functions. In the gribble (*Limnoria lignorum*) which is an isopod, the mandibles are specialized, with one like a rasp, the other like a file. The species causes extensive damage by boring marine timbers. The mandibles of many insects cross asymmetrically giving a scissor action (scissors are asymmetrical). The wings in many orthoptera are asymmetrically adapted for sound production. We have seen in Chapter 3 that many parasites are asymmetrically adapted when they inhabit cramped habitats. In some very advanced parasites there may be total loss of symmetry, e.g. the parasitic barnacle *Sacculina*. Hermit crabs are beautifully adapted for inhabiting the dextral shells of gastropods. Another animal which has acquired a spiral twist is *Phascolion* (Sipunculoida) which inhabits discarded turret shells, e.g. *Turritella*. The asymmetrical beak of the crossbill is adapted for extracting seeds from conifer cones.

Pattern asymmetry is useful for camouflage, e.g. viper, flatfish, larvae of the fly *Thaumalea*. In some owls the ear asymmetry is used for pinpointing prey by sounds in the dark, a most useful adaptation to nocturnal habits. Some examples of spiral fibre systems are used in locomotion to replace a set of muscles, thereby saving energy. In many situations fibre orientation is adapted for improving mechanical strength (Chapter 4). Brain asymmetry in man has permitted each side of the brain to specialize in different functions. A nice example of chemical asymmetry is seen in the oothecal glands which secrete the case around the eggs in preying mantids and cockroaches. The large left gland produces structural proteins, the glucoside of a tanning agent precursor and phenoloxidase. The smaller right gland produces a glucosidase. When the two mix, the glucosidase liberates the tanning agent precursor from the glucose which was attached to it. This precursor is now available for oxidation to a very reactive cross-linking agent by the phenoloxidase. This agent then cross-links the structural proteins to harden the egg case. Thus the differentiation of compartments functions to prevent premature hardening before the appropriate time. It is rather like types of glue which harden on mixing two different components.

7 How is Asymmetry Achieved?

We have seen in the previous chapters that examples of animal asymmetry are numerous and that they involve many levels of structure. Intuitively, therefore, we would not expect a unique answer to the title of this chapter, which is meant to provoke interest in an unsolved question.

7.1 Bias in asymmetry

In some examples there is an equal chance of an asymmetrical structure lying on either the left or the right side of the body (e.g. the primitive flatfish, *Psettodes*). In the fiddler crab, *Uca* or the lobster, *Homarus*, the deciding factor dictating the position of the large claw is environmental, by chance injury. It is not difficult to picture a system of growth-controlling morphogens, produced in the claw appendage, which are normally balanced across the body. Removal of one claw could then upset the balance and lead to asymmetrical growth.

What are more interesting are the systems where there is *not* an equal chance of left or right asymmetry. Table 4 gives some examples of what are termed here *bias in asymmetry*, expressed as percentages, and covering a wide range of animals. It is clear that some species are able to select left from right during development, and in some cases to make a correct choice nearly every time. This is paradoxical because the gene pool is supposed to be the same on both sides of the body. Furthermore, environmentally induced bias would be expected to fall close to 50%. Yet in some cases asymmetrical bias varies in a geographical manner (Table 5). It might be well to point out, however, that the probability of a departure from a 50 : 50 expectation even in the extreme case of Santander is about 0.04. We have already seen (section 5.3) that the frequency of gynandromorphy in insects can also vary geographically. These then are the problems.

A trivial explanation of bias could be that there is a difference in mortality rate between left and right individuals. There is some evidence for this in a Baltic flounder, *Platichthys flesus*, since sinistrality is higher in young fish (55%) than in adults (31%), indicating reduced vitality in reversed individuals. But extensive experiments with huge numbers of snails, *Limnaea peregra*, did not support this idea (BOYCOTT *et al.*, 1931).

Certain inhibitors of the enzyme carbonic anhydrase (e.g. acetazolamide and dichlorphenamide) induce specific deformity in the right forelimb of rats. Such mutagens can thus induce asymmetrical bias.

Table 4   Examples of left or right bias in asymmetry.

Examples	Species	%
Sinistral shell	*Limnaea peregra*[1]	0.001
Sinistral shell	*Buccinum*[1]	0.1
Large L. chela	*Homarus americanus*[2]	53
L/R wing folding	*Gryllotalpa gryllotalpa*[3]	0.05
L/R wing folding	*Acheta domesticus*[3]	0.05
L/R wing folding	*Acheta assimilis*[3]	4–10
L/R wing folding	*Acheta pennsylvanicus*[3]	4–10
L/R wing folding	*Acheta rubens*[3]	4–10
L/R wing folding	*Teleogryllus commodus*[3]	10
L. abdominal torsion	*Clunio marinus*[3]	48–55
Eyes on L. side	*Solea*[4]	0·001
Eyes on L. side	*Pleuronectes platessa*[4]	0.001
Eyes on L. side	*Psettodes*[4]	50
Gut reversal	*Cottus gobio*[4]	0.2
More fin rays in R. pectoral fin	*Leptocottus armatus*[4]	60–81
Gut reversal	*Triton*[5]	2–3
L. gular plate larger	*Gopherus agassizii*[6]	90
Lower mandible crosses L.	*Loxia curvirostra*[7]	58
Extra digit on L. foot	*Gallus domesticus*[7]	90
L. handedness	*Homo sapiens*[8]	5

Key: 1, Mollusca; 2, Crustacea; 3, Insecta; 4, Osteichthyes; 5, Amphibia; 6, Reptilia; 7, Aves; 8, Mammalia.

7.2   Enantiomorphic forms

It is clear that the left and right forms of snails are established by the orientation of mitotic spindles (Fig. 5.1) during early cleavage. The result can first be seen at the four cell stage when the cells are arranged tetrahedrally. If two of the cells exchange position, mirror imaged snails result. Here both spindles and cell walls are involved. But in the Protozoan *Zoothamnium* (section 3.2) cell wall division is not required for setting up the branching pattern in left and right colonies. Hence asymmetry can be controlled at the intracellular level without the need for communication between cells.

Reversal in flatfish is not usually a complete reversal such as is seen in sinistral gastropods, because it does not involve *situs inversus* of the viscera. One exception is, however, known in *Tanakius kitaharae* (HUBBS and HUBBS, 1945). This independence between levels of asymmetry argues for a hierarchy of control mechanisms. In a few cases, left and right forms are environmentally controlled, e.g. injury to claws in lobsters, and pressure of left or right gill chambers of host on *Bopyrus*. It is thought that sense of shell coiling in the Foraminiferan *Globorotalia truncatulinoides* changes with

Table 5 Geographical variation in asymmetry bias.

A *Torsion direction of the male abdomen in* Clunio marinus *(Dordel 1973)*

Locality	n	L. torsion	R. torsion
Tromsø (Norway)	324	155	169
Heligoland	379	189	190
Quiberon (Brittany)	448	228	220
St Jean de Luz	453	242	211
Santander (N. Spain)	435	240	195

B *Reversal (sinistrality) in Pacific flounder,* Platichthys stellatus *(Hubbs and Kuronuma, 1942)*

Locality	Percentage
Off N.W. America	49–60
Off Alaska	68
Off Japan	100

water temperature. Enantiomorphs are a special example of polymorphism.

7.3 Differentiation

Many of the examples of asymmetry concern the asymmetrical differentiation of homologous cells on opposite sides of the body. The problem falls into a general category: how do the two daughter cells of a mitosis produce different products? Familiar examples are the asymmetrical divisions in oogenesis into oocytes and polar bodies; and the formation from one insect epidermal cell by two mitoses of a neuron, a cell forming a sheath around it, a cell forming the sensory hair which the nerve cell supplies, and a cell forming a socket for the hair.

One way in which asymmetrical differentiation can come about is by the unequal allocation of chromosomes during cell division, e.g. gynandromorph formation in insects (section 5.3). Its occurrence is, however, too erratic to explain the bias figures for other systems.

Again, a subtle difference in micro-environment could produce asymmetrical differentiation. The cells at the bases of mammalian gut villi divide to produce one primitive cell like the parent (which retains mitotic ability), and one further differentiated cell which cannot divide. The cell division is thought to be symmetrical but the latter cell is pushed out into the villus, where it encounters a different micro-environment which favours differentiation. If the mechanism goes wrong, with both daughter cells remaining undifferentiated and still able to divide, the cell population would rise (irrespective of the rate of mitosis) and a cancer could result.

A theory has recently been proposed to account for asymmetrical differentiative division from a symmetrical parent cell in a uniform environment, by non-homogeneous chemical perturbations. It does not, however, explain how bias can arise such that a left cell becomes differentiated in one way, whereas a right cell differentiates in another.

7.4   Genetics and asymmetry

It is known that sinistrality and dextrality in *Limnaea* are controlled genetically (section 5.2). Also, hybrids produced by mating sinistral *Platichthys stellatus* with dextral *Kareius bicoloratus*, comprised 14 sinistral and 13 dextral offspring. Therefore reversal is genetically determined. In mice, the luxate mutation affects one side rather than the other.

7.5   Cell gradients

Cell polarity gradients can affect asymmetry in two ways. In a species which normally has bilateral symmetry, asymmetry can arise if something disturbs the gradients in the odd individual. Again, there may be a cell polarity gradient running from left to right across the body in an animal in which asymmetry is normal (section 5.8). If this is so, we still have to ask how the gradient, which is itself asymmetrical, is set up in the first place. This argues for asymmetry in the original cell before cleavage.

7.6   Self-assembly

We have seen that there is reasonable evidence that the asymmetrical chitin helicoid in arthropod cuticles, and the asymmetrical collagen system in vertebrate eyes, may arise by self-assembly. In this case the chemistry of the system would determine the sense of rotation. The evidence is circumstantial at present and needs chemical documentation.

7.7   Conclusions

Finally, we can try to synthesize several observations. (1) Mitosis normally gives rise to two daughter cells with equal chromosomes. (2) Unicellular organisms can show asymmetry (e.g. *Paramecium*, and the branching stalks of *Zoothamnium*). (3) There is asymmetry in the cytoplasm of *Limnaea* oocytes. (4) The planes of mitotic spindles in developing *Limnaea* subsequently determine cleavage sense, and the sense of the resulting snail. (5) Asymmetry is usually under genetic control.

We can tentatively conclude that asymmetry resides fundamentally in the cytoplasm of a cell, and that this has been determined precisely by the nucleus. Subsequent division leads to duplication of chromosomes but to an inequal distribution of the asymmetrical components in the cytoplasm. The latter presumably possess the power of self-replication, to prevent their dilution with subsequent divisions. It would be of interest to try to relate the time when a new level of asymmetry is being established, with time of transcription of messenger RNA from DNA.

If asymmetry exists at the cellular level, cell polarity gradients running asymmetrically across an organism may then perhaps be envisaged as the alignment of several asymmetrical cells as if they were magnets.

Appendix: Projects

Beginning with simpler projects, it would be instructive to make a collection of asymmetrical animals, using this book as a guide. The various asymmetries which they show could then be related to function or development. Fieldwork could also be involved in measurements of left/right bias percentages in several asymmetrical groups. Examples are abdominal torsion in male flies; sinistrality and dextrality in snails; reversal in flatfishes (most appropriate for those living near fishing harbours); owl skulls; possible bias in bivalve mollusc hinges, which has not been tackled; sense of direction of human hair whorls (highly suitable for schools): insect mandible asymmetry; insect wing folding preference (here a light trap for moths would be an ideal collecting method, prior to analysis and release). For the behaviourists, investigation of spiral locomotion in humans is clearly suitable for schools, and simple observations on asymmetry during tadpole metamorphosis are also easily performed. Building molecular models of isomers of amino acids and relating them to the sense of twist of an α-helix is most instructive. Liquid crystal kits for making asymmetrical liquid crystals can be obtained from Vari-light Corporation, Cincinnati, Ohio, 45242 as well as from Peboc Ltd., Industrial Estate, Llangefni, Anglesey, Wales, and the necessary cheap optical accessories for observing them are obtainable from Polarizers U.K. Ltd., Lincoln Road, Creffex Industrial Estate, High Wycombe, Bucks.

More advanced morphological projects include measurements of sense of layer rotation in various vertebrate eyes using the Nomarski interference method. It is possible that many epidermal basement membranes may show layer rotation when this method is used. There is a chance using *Zoothamnium* to try to find possible enantiomorphy between individuals at the cell organelle level, by electron microscopy. Whale researchers may like to follow up the idea that skull asymmetry in carnivorous species improves hearing direction.

The method of mapping neurons in gynandromorphic insect mosaics offers a valuable complement to methods already in use. For biochemists there is the challenge to isolate the morphogens controlling asymmetry of bud reversal in *Botryllus* and gut in Amphibia.

References

Space forbids the inclusion of all references cited in the text. However, these may be located via *Zoological Record* or *Biological Abstracts*. Some general references are given below.

AGENO, M. (1972). On molecular asymmetry in living organisms. *J. theor. Biol.*, **37**, 187–92

BOULIGAND, Y. (1972). Twisted fibrous arrangements in biological materials and cholesteric mesophases. *Tissue and Cell*, **4**, 189–217

BOYCOTT, A. E., DIVER, C., GARSTANG, S. L. and TURNER, F. M. (1931), The inheritance of sinistrality in *Limnaea peregra* (Mollusca, Pulmonata). *Phil. Trans. B.*, **219**, 51–131

FRANK, F. C. (1953). On spontaneous asymmetric synthesis. *Biochim. Biophys. Acta.*, **11**, 459–63

GARAY, A. S. (1968). Origin and role of optical isomery in life. *Nature, Lond.*, **219**, 338–40

GARDNER, M. (1970). *The Ambidextrous Universe. Left, right and the fall of parity.* Pelican Books

GAZZANIGA, M. S. (1967). The split brain in man. *Scient. Amer.*, **217** (2), 24–9

HUBBS, C. L. and HUBBS, L. C. (1945). Bilateral asymmetry and bilateral variation in fishes. *Pap. Mich. Acad. Sci., Arts and Letters*, **30**, 229–310

HUXLEY, J. and DE BEER, G. R. (1963). *The Elements of Experimental Embryology.* Hafner, New York. (Especially Chapter 4)

NEVILLE, A. C. (1970). Cuticle ultrastructure in relation to the whole insect. In *Insect Ultrastructure*. Edited by A. C. Neville. *Symp. R. ent. Soc. Lond.*, **5**, 17–39

NEVILLE, A. C. (1975). *Biology of the Arthropod Cuticle.* Springer-Verlag, Berlin.

NORBERG, Å. (1968) Physical factors in directional hearing in *Aegolius funereus* (Linné) (Strigiformes), with special reference to the significance of the asymmetry of the external ears. *Arkiv. Zool.*, **20**, 181–204

OPPENHEIMER, J. M. (1974), Asymmetry revisited. *Am. Zool.*, **14**, 867–79

SCHAEFFER, A. A. (1928). Spiral movement in man. *J. Morph. Physiol.*, **45**, 293–398

WILSON, D. M. (1968). Inherent asymmetry and reflex modulation of the locust flight motor pattern. *J. Exp. Biol.*, **48**, 631–41

B7